W0042338

Monographs in Visual Communication

David F. Rogers
Editor

Monographs in Visual Communication

Hall: Illumination and Color in Computer Generated Imagery
von Bechtolsheim : TEX in Practice, Volume I: Basics
von Bechtolsheim : TEX in Practice, Volume II: Paragraphs, Math, and Fonts
von Bechtolsheim : TEX in Practice, Volume III: Tokens, Macros
von Bechtolsheim : TEX in Practice, Volume IV: Output Routines, Tables

Stephan von Bechtolsheim

TEX in Practice

Volume II: Paragraphs, Math and Fonts

Springer-Verlag
New York Berlin Heidelberg London
Paris Tokyo Hong Kong Barcelona Budapest

Stephan von Bechtolsheim
c/o Springer-Verlag New York
175 Fifth Avenue
New York, NY 10010
USA

Library of Congress Cataloging-in-Publication Data
Bechtolsheim, Stephan von
TEX in practice / Stephan von Bechtolsheim
p. cm. (Monographs in visual communication)
On t.p. E is subscript
Includes bibliographical references and indices.
Contents: v. 1. Basics v. 2. Math and fonts v. 3. Tokens, macros v. 4. Output routines, tables.
ISBN-13:978-1-4613-9141-8 e-ISBN-13:978-1-4613-9139-5
DOI: 10.1007/978-1-4613-9139-5
1. TeX (computer system) 2. Computerized typesetting 3. Mathematics printing.
I. Title. II. Series.
Z253.4.T47B4 1992
686.2"2544 dc20 90-10034

Printed on acid-free paper.

Production supervised by Kenneth Dreyhaupt, manufacturing coordinated by Vincent Scelta.
Photocomposed copy prepared from the author's TEX files.

9 8 7 6 5 4 3 2 1

ISBN-13:978-1-4613-9141-8

To my wife Alice and my son Benjamin
Who patiently waited for me to get done.

Foreword

You might well wonder why *TeX in Practice* is a part of the Monographs in Visualization series. However, if you really think about typesetting, especially fine typesetting, you soon realize that in large part it is a *visual art* as well as a science. TeX's algorithms produce in almost all cases aesthetic results of the highest quality. On the other hand, occasionally one may want to insert some additional space before a subscript or superscript, or one may want to adjust the vertical spacing in a fraction. Fortunately Donald Knuth, the author of TeX, allows one to program such corrections easily where needed.

The four volumes of Stephan von Bechtolsheim's long awaited *TeX in Practice* present a comprehensive view of TeX. His thorough discussion of each aspect of TeX is liberally laced with cogent illustrative examples. Many of these examples represent complete, ready to use macros that enhance the capabilities of TeX. These examples are of particular interest to both the typesetter and the TeX programmer. The typesetter can often solve an immediate problem by either using one of the examples directly or by making minor changes to adapt it to the problem at hand. The TeX programmer can use the examples, along with Stephan's detailed discussion, to increase both the depth and breadth of his or her knowledge of TeX. The value of the text is further enhanced by Stephan's concerted effort to explain the reasoning behind each topic or example. In many cases, he details the inner workings of TeX's processing of the example.

Stephan is to be congratulated on producing a work of fundamental and lasting value to the TeX and publishing community.

David F. Rogers
Annapolis, Maryland
May 1993

Preface

Although you only have one volume in front of you, writing four volumes and 1600 pages on a single subject needs some form of justification. And then on the other hand, why write even more?! Can't, at least, the preface of something that long be short?!

Very well, so let's keep it short. It is my sincere hope that the series "TEX in Practice" will be useful for your own TEX work. But *please*, before you get started, read the "Notes on 'TEX in Practice'," because it instructs you how to use this series. You will find these notes on pages xxxi–xl.

The second volume of this series deals with paragraphs, math mode and fonts in TEX. Paragraphs, in particular, are a very basic entity in TEX, and you will be amazed to see what different paragraph shapes you can typeset in TEX. TEX is famous for the typesetting of mathematical equations and we will look at that in this volume too.

Stephan von Bechtolsheim

Arlington Heights, Illinois
May 1993

Table of Contents

List of Figures

List of Tables

Acknowledgements

A long list of people is what a long book deserves. So let's get started.

My editors, Kim Miller (who did the bulk of the work) as well as Kris Schlenker and Judith Lewis, tried patiently to make this series sound like "real English." If it doesn't sound right always, blame me (a quote from one of my students: "by now you are probably *bilingually illiterate*"). Last minute corrections on my part may also have introduced errors.

At Purdue University Chris Hoffman and Bob Lynch (Computer Sciences Department), Mark Senn (Statistics Department; by the way he is the person who brought TeX to Purdue), and Steve Samuels (also Statistics Department) helped; there was also Ed Ramsey (National Pesticide Center of Purdue University) who let me use his SPARCStation to process this series. Clients of mine such as Siemens-Nixdorf Corporation (Jörg Heinemann, Joachim Haenel, Bernd Stümke), Jörg Steffenhagen, of Munich, West Germany, and Technical Typesetting, Inc. (Bill Taylor and Harry Kirk) located in Philadelphia allowed me to publish macros I had developed for them. Gerhard Rossbach, Mark Hall, and Rüdiger Gebauer of Springer-Verlag all waited patiently for me to finish my series. Dave Rogers (United States Naval Academy, Anapolis) gave some valuable hints. Chris Thompson (Cambridge University Computing Service) and Thomas Reid (Computing Services Center, Texas A&M University) provided some input too. A.C. Conrad of the Menil Foundation in Houston hired me as a consultant. Some of the TeX problems I solved for him found their way into this series. I used Oliver Schoett's idea (Institut für Informatik, TU München, München (or should I have written "Munich"), Federal Republic of Germany), designing macro `\PrintHyphens`. One of the tables in the table chapters was submitted to me by Cynthia Rodriguez, Computer Center, University of Illinois at Chicago. Michael Doob, University of Manitoba in Canada, helped with a problem in the output routine chapters. Kabelschacht's article (Kabelschacht (1987)) was used in a chapter in this series.

To thank Donald E. Knuth, the author of TeX and METAFONT, warrants a separate paragraph, in particular, because he replied to my mail instantaneously even after he had announced that he would not be involved in the further development of TeX. Not only that, *he* provided TeX to all of us! Thank you *very much.*

Leslie Lamport also deserves special mentioning. When I started to write this series, there were plenty of things I did not understand. For instance, verbatim mode was one of the things I initially simply ignored. What I did was rather straightforward: I started to write this series using LaTeX. As time proceeded I replaced one part of LaTeX after another by my own macros. You will probably also recognize some macro names in this series which remind you of LaTeX macro names such as `\Section` (written `\section` in LaTeX) and `\normalsize` (called the same in LaTeX). Even now some of the diagrams are done with LaTeX's picture environment. Other things I stole / borrowed / adopted include the definition of macro `\frac`, the idea of a font selection scheme which separates size and type face, etc.

Victor Eijkhout, Center for Supercomputing Research and Development, University of Illinois at Urbana-Champaign, also deserves his own paragraph: he actually went through the whole series. Maybe he did not read every page, but he certainly read *almost* every page. There were a lot of little things he pointed out to me.

Of course, what would be a series on TeX without mentioning Barbara Beeton of the American Mathematical Society in Providence, Rhode Island, who helped me with a lot of small details. Ray Goucher of the TeX Users Group, who was a good boss while I taught classes for the TeX Users Group, deserves acknowledgement too.

Addison-Wesley gave permission to reprint and distribute (in modified form) some macros of the TeXbook (see 16.8, p. 297). Penguin Books gave permission to use Shakespeare (1605) for examples in Chapter 10 (this material is in the public domain, but I thought you might want to know where I got it from).

The Free Software Foundation must be mentioned too, because it provides the GNU Emacs to which I am addicted by now. GNU Emacs is the only editor you want to use after you have used it a couple of times. It certainly has done wonders for me. I designed my own "TeX-mode" and that made the tedious task of editing hundreds of pages much more tolerable.

Finally my current employer (Datalogics Incorporated in Chicago, a subsidiary of Frame Technology of San Jose, California) allowed me to use one of their printers for printing the last n revisions of this series ($n < 10$).

Trademarks

The following trademarks are used in this series.

TEX is a trademark of the American Mathematical Society; METAFONT is a trademark of Addison-Wesley Publishing Company; POSTSCRIPT is a trademark of Adobe Systems Incorporated; UNIX is a registered trademark of AT&T; DEC, VAX, and VMS are registered trademarks of Digital Equipment Cooperation; Apple LaserWriter, Macintosh are trademarks of Apple, Inc.; IBM is a registered trademark of International Business Machines Corporation; HP LaserJet is a trademark of Hewlett Packard; SUN is a trademark of Sun Micro Systems, Inc.; Mercedes is a registered trademark of Mercedes-Benz AG, Stuttgart, Federal Republic of Germany; SPARC is a registered trademark of SPARC International, Inc., licensed exclusively to Sun Microsystems, Inc.; SPARCStation is a registered trademark of Sun Microsystems, Inc.; Porsche is a registered trademark of Dr. Ing. h.c. F. Porsche AG.

Other brand or product names are the trademarks or registered trademarks of their respective holders.

A Note About the Dedication

This dedication needs little explanation. Let me just say this: writing a book seems to always take longer than anticipated (if there was anything anticipated in the first place), and that strains the nerves of one's family.

Stephan von Bechtolsheim

Arlington Heights, Illinois
May 1993

General Notes on "TEX in Practice"

This part of the book contains some general comments about the series "TEX in Practice." It is reprinted in the beginning of every volume of this series and provides kind of a "general introduction and series-wide table of contents." I urge you to read the following material carefully.

1 Some Friendly Advice Up Front Or What To Do If You Hate to Read Lots of Material

If you hate to read lots of material you are in trouble, obviously. But here are some hints anyway (who knows whether you even read this paragraph ...):

- Continue reading this part, "General Notes on 'TEX in Practice'," please, please.
- Quickly look through the tables of contents of all four volumes. I suggest limiting yourself to reading the title of all chapters and sections, but to ignore subsections and further subdivisions.
- Read the overview of all chapters on p. xxxiii.

Thank you!

2 Terminology, Conventions, Other General Remarks

Let me make some general comments about the terminology used in "TEX in Practice" and some conventions I followed.

1. The series "TEX in Practice" is subdivided into *four volumes.* Each volume is identified by a *capital Roman numeral.* Each volume is subdivided into *chapters.* Chapters are identified by *arabic numerals* and numbered across volumes. Each chapter is further subdivided into *sections* (the second section of chapter 12 would be numbered 12.2), and sections, in turn, are divided into *subsections* (the fourth subsection of section 12.2 would be numbered 12.2.4). There are also *subsubsections,* which are numbered accordingly.

   Note that in cross-references to sections, subsections, and subsubsections the words section, subsection or subsubsection respectively are dropped. So a reference to Section 12.3 might read "see 12.3, p. 23, for further details." References to other entities like chapters, figures or tables are "spelled out" that is a reference to a figure might read "Fig. 3.4."
2. Arabic page numbers are used with the exception of the preliminaries of each volume in which Roman page numbers are used. The page numbering starts at page 1 in every volume.

   In references to page numbers, the volume number precedes the page number; in references to the *current* volume the volume number is dropped. For instance, "p. 23" refers to page 23 in the current volume, and "p. I-12" refers to page 12 of the first volume (in any volume other than the first).
3. The table of contents of each volume lists all chapters, sections and subsections of the volume, but not the subsubsections.
4. "TEX in Practice" tries to give a comprehensive and detailed view of TEX, but it does *not* claim to be complete. For very tricky questions I suggest that you first consult this series, but then also look into the TEXbook. I hope that reading this series will make understanding the TEXbook much easier, but I skip over some of the more bizarre details (which the TEXbook does not).
5. In this series I need to occasionally refer to the series itself. I do so using the expression "this series."
6. In general in this series I refer to TEX 3.0. The extensions implemented in TEX 3.0 are described in Knuth (1990). Note that TEXbooks published prior to 1990 do not contain a description of these extensions. The 3.0 extensions are minor, but nevertheless important. In some instances I will identify those differences by a short phrase such as "a TEX 3.0 feature" to identify a TEX 3.0 extension (*not all* such extensions are identified as such).

3 The Intended Audience of "TEX in Practice"

Let me now discuss whom this series was written for. The more of the following statements apply to you, the better:

- You want to learn TEX thoroughly because you have to typeset complicated documents. Your text might include mathematical formulas and tables; both areas are very well covered by TEX.

- You are a curious person: when you use a tool (and TEX, of course, is a typesetting tool), then you wish to *understand* the tool. You do *not* only want to apply it.
- You believe that having lots of examples will allow you to understand something better. This series contains plenty of examples.
- You think that designing your own macros is something you may want to do, now or in future, because none of the existing macro packages fulfill your needs precisely. This series should come in handy, because it already contains the basics of a fairly sophisticated macro package plus it contains many basic utility type of macro definitions.
- You recognize that WYSIWYG (What-You-See-Is-What-You-get) systems, where your typeset output is immediately displayed on the screen, are limited in their power in that they cannot be easily adopted to specialized needs.

  Also note, that WYSIWYG systems cannot solve the real problem of writing, that is to present thoughts in an orderly and organized fashion, either. Owning a Porsche doesn't make you necessarily a good driver (though probably a happy one), if you know what I mean.

4 A Brief Overview of "TEX in Practice"

The series "TEX in Practice" is divided into four volumes. An overview of each volume follows.

1. Volume I:

   - Chapter 1 explains the existing relationship in the complexities of problems and tools to solve these problems. It explains that the complexity of TEX is a direct consequence of its power.
   - Chapter 2 provides a general introduction into TEX. It explains the basics of entering data into TEX and processing a document.
   - Chapter 3 begins the discusses of registers concentrating on counter registers primarily. Generic counter macros are discussed and an overview of all TEX's parameters is given.
   - Chapter 4 continues the register discussion looking at dimension registers and box registers. Also glue registers are discussed
   - Chapter 5 discusses glue (a very important concept of TEX), rules and leaders.
   - Chapter 6 discusses horizontal boxes. This is important in itself and also forms the basis of the discussion of the following two chapters. We also discuss how to do tables using hboxes.
   - Chapter 7 discusses vertical boxes, their dimensions, and different types line spacing and struts.
   - Chapter 8 discusses more advanced concepts relating to vertical boxes such as `\vsplit` and the shifting of reference points of vboxes.

- Chapter 9 discusses boxes and rules and introduces macros to draw rules around boxes. Boxes with rules are a very useful debugging feature of TeX.

2. Volume II:

- Chapter 10 discusses how paragraphs are built. You find the very basics of typesetting text in this chapter.
- Chapter 11 continues the discussion of the previous chapter and shows how more complex paragraph shapes are typeset.
- Chapter 12 concludes the discussion of the typesetting of paragraphs.
- Chapter 13 discusses the basics of typesetting mathematical formulas.
- Chapter 14 continues the discussion of typesetting mathematical formulas.
- Chapter 15 discusses fonts. This discussion includes an introduction to the Computer Modern fonts and how special fonts can be used in TeX.
- Chapter 16 continues the discussion of fonts. A major point in this chapter is the proper organization of fonts.
- Chapter 17 deals with the "environment" of TeX. This includes the discussion of METAFONT and the WEB system.

3. Volume III:

- Chapter 18 discusses tokens, grouping and category codes. You will also find an explanation of verbatim modes.
- Chapter 19 continues the discussion of the previous chapter explaining cross-referencing, list macros, and the use of token registers.
- Chapter 20 discusses token lists and arrays.
- Chapter 21 explains the basics of macro definitions, based on `\def` with delimited and undelimited parameters.
- Chapter 22 continues the discussion with macro definitions based on `\edef`, nested macro definitions and goes into more details about macro arguments.
- Chapter 23 discusses `\let`, `\futurelet` and `\afterassignment`.
- Chapter 24 discusses `\expandafter` and its applications as well as `\the`.
- Chapter 25 discusses conditionals.
- Chapter 26 discusses date and calendar related macros.
- Chapter 27 discusses more additional macro related issues such as recursion, environments and error handling.
- Chapter 28 discusses input and output. This includes a discussion of delayed and immediate writes, writing of table of contents files, and literal writing.
- Chapter 29 continues the preceding chapter discussing index file and verbatim file macros. It also discusses table of contents and endnotes generation.
- Chapter 30 discusses how the partial processing of a document can be administered (one document's text is stored in separate files which also

should be processed separately).

- Chapter 31 continues the discussion of the preceding macro. It in particular offers some rather sophisticated and interesting cross-reference macros.

4. Volume IV:

- Chapter 32 begins the discussion of tables typeset with TEX's `\halign` instruction.
- Chapter 33 also discusses tables, including vertical spacing, struts and rules in tables.
- Chapter 34 discusses the centering of tables, tables and paragraphs and preamble related macros.
- Chapter 35 discusses numerical computations in tables, splitting tables, `\valign` and double tables.
- Chapter 36 discusses the determination of page breaks by TEX.
- Chapter 37 discusses the basics of output routines, and introduces concepts like logical and physical pages.
- Chapter 38 presents some simple output routines.
- Chapter 39 discusses the output routine of the plain format and variations of it.
- Chapter 40 discusses output routines with insertions.
- Chapter 41 discusses double column output routines.

Each volume furthermore contains two prefaces, the first one is common to all volumes, the second one is a preface particular to the specific volume. The bibliography and the acknowledgements are identical in all volumes as is the "General Notes on 'TEX in Practice,'" the part you are currently reading.

One other brief point: I found it *extremely* difficult to present TEX in a linear order, without forward references. There is of course the possibility of making two or three passes over a subject such as TEX, each time with increasing difficulty. If I had done that, this series would have become even longer. At least in the approach I took I found it unavoidable and regrettable to have a fair number of forward references.

5 Using the Macros Presented in "TEX in Practice"

I hope that the macros presented in this series are useful, and that you will actually apply these macros. Here are some general comments about these macros, how they should be used, and so forth.

1. All macro source files presented in this series have the file extension `.tip` for "TEX in Practice." The file extension `.tex` is *not* used by any of the macro

source files. Some of the example source files use file extension `.tex`.

I *ask* everyone to *not* use the file extension `.tip` for any files other than the files of this series. I also ask you to *not* change any of the source code files of this series. If you need to change any of the published macro source code files (which you are encouraged to do, because one of the main purposes of this series is to use all the presented macros), then please *rename* the file you are using, and please also *change the file extension* at that point.

2. There are three classes of TeX source files presented in this series:

   - *Published* files. Those are those files, which are available to you in machine readable form. The published macro source files are identified by a $\mathcal{P}$ in the line preceding the beginning of the source code listing. The file names of these files are at most 8 characters long, a restriction which is probably useful for the PC environment.

     Here is an example of a published source code file.

     $\mathcal{P}$ • `publish.tip` •

```
\def\SampleMacro{%
    ... Sample of a macro which is published on a diskette.
}
```

     • End of `publish.tip` •

   - Most of the published files are pre-loaded as part of the `texip` format which is defined in 31.3, p. III-612. Those macros *need not* be loaded, *if* the format defined at the given reference is used. Those macros are marked by a $\mathcal{P}'$.

     See item 13, p. xxxix, for some additional information.

   - *Unpublished* files. In most cases unpublished files are example files, which apply one or more of the published files. These files are obviously *not* marked by $\mathcal{P}$.

3. The file names of those macro source files, which were designed particularly for the publication of this series, start with "`ts-`," which stands for "TeX in Practice Series."

4. To get your own copy of the published macro source files contact one of the following places:

   (a) The TeX Users Group (see further below for the address).
   (b) Your favorite TeX supplier.
   (c) The standard places for TeX source code on the networks, if you are hooked up to one of the national or international networks.

   If you acquire those macros commercially, you must *not* be charged any fee other than a nominal fee for media and shipping (the same way you do not have to pay extra for the plain format or LaTeX macro source files). One more request: if you modify a macro, you are required to rename it, even if the modification is minor. With the exception of the conditions laid down in this paragraph you can do anything reasonable with these macros you want.

5. The following additional information applies, in particular, if you retrieve

the macro source files using a network (I try to establish a convention here and I hope everyone will follow it).

(a) The macros will be stored in a separate subdirectory called **`texip`** (all lower case).
(b) With the macros comes a file **`readme.tip`** which you should read first. This file contains the following information:

    i. A release date of all macros.
    ii. A version number for all the macros together. This version number is a "combined" version number of all individual macro source files: it is the maximum version number of any individual source file's version number.
    iii. A short history of the changes made to any of the macro source files.
    iv. A complete list of *all macro source files* and their version numbers listed in *alphabetical order.*

6. Every macro source file starts out with the same standardized text, which reads as follows (the last four lines are, of course, different from file to file, also note that the file name such as **`reg2.TEX`** as well as the line numbers refer to my own source code files and are therefore of secondary interest for the user of the macros in this series):

```
1  % This macro source file is from the four volume series
2  % "TeX in Practice" by Stephan von Bechtolsheim, published
3  % 1993 by Springer-Verlag, New York.
4  % Copyright 1993 Stephan von Bechtolsheim.
5  % No warranty or liability is assumed.
6  % This macro may be copied freely if no fees other than
7  % media cost or shipping charges are charged and as long
8  % as this copyright and the following source code itself
9  % is not changed. Please see the series for further information.
10 %
11 % Version 1.0
12 % Date: May 1, 1993
13 %
14 %
15 % This source code is documented in 4.1.11, p. I-91.
16 % Original source in file "reg2.TEX", starting line 587.
17 \wlog{L: "absdimen.tip" ["reg2.TEX," l. 587, p. I-91]}%
18 % This file DOES belong to format "texip."
```

The previous text is 18 lines long. The text is *not* reprinted in the series itself anywhere else, in particular not with every source code listing. Therefore, to synchronize the line numbers of the source code as listed in this series and the source code of the `.tip` files, all *listings* of published source code files start with line 19, rather than line 1.

7. To *use* any of these macros, *always* use the following approach:

    (a) First load file `inputd.tip` using `\input` (only only):

```
\input inputd.tip
```

(b) Next load all macro files you want to load by calling the macro `\InputD` (do *not* use `\input` to load any `.tip` macro source files with the exception of `inputd.tip`). The `\InputD` macro has one parameter, `#1`, the name of a source code file to load.

For instance, to load file `box-mac.tip`, enter `\InputD{box-mac.tip}`. Do *not* omit the file extension `tip`. The beginning of your TeX source file would now look as follows:

```
\input inputd.tip
\InputD{box-mac.tex}
```

More `\InputD` calls could follow at this point.

8. Note that this series defines a format which has most of the relevant macros defined in this series pre-loaded. See 31.3, p. III-612, for details.
9. The series contains the reprint of numerous log files. The version of the TeX program I was running for the processing of this series was changed to print shortened log file lines to accommodate the width of the pages in this series.
10. In case you use LaTeX with any of my macros, you should not have any problems. Enter the instruction to load `inputd.tip` and the subsequent calls of the input macro `\InputD` *after* the `\documentstyle` command and *before* `\begin{document}`.

    Here is a brief example of what your LaTeX source code file might look like:

```
\documentstyle{article}
\input inputd.tip
\InputD{box-mac.tip}
\InputD{...}
...                   % Your macros and initializations go here.
\begin{document}
    ...
\end{document}
```

Two more remarks about using these macros with LaTeX:

(a) Be aware that LaTeX uses `\everypar`. Therefore, if you use any of the macros from this series which change `\everypar`, you must be aware of potential conflicts.

(b) The plain format and LaTeX both have a `\line` macro with two totally different and unrelated meanings.

11. Remember, this series teaches you how to use TeX. The macros do *not* form a complete format including style files. In a typical application you will load macros and then write yourself additional macros. This series is like a cookbook: you find lots of recipes, but you still need to get out the utensils yourself, decide on the menu and clean up later. If you need something ready to use I suggest using LaTeX.
12. I cannot assume any *liability* for the macros. But I did my best to provide error free code. Please contact me in case you find any errors (in the source

code of the presented macros or otherwise in this series). The next Section contains my address.

13. If loading individual macros, etc., seems to be a little cumbersome to you, we provide to you (as part of the macros) a special file called `texip-ex.tip` (the file name stands for `texip.tip`, expanded). Here the word expanded means that this is *one* macro source file, which (with the exception of `plain.tex`) contains *all* macro source files marked with $\mathcal{P}'$.

    Whether you run `initex` on `texip.tip` (which loads another 200 or so files) or run `initex` directly on `texip-ex.tip` does not make any difference as far as the end result is concerned. It's just easier in that if you just pick up `texip-ex.tip` from somewhere then you don't need to pick up other files, at least initially.

    I suggest that if you use `texip-ex.tip` that you rename the resulting `.fmt` file from `texip-ex.fmt` to `texip.fmt`.

6 Contacting the Author

If you wish to get in touch with me, please write to the following address. I am interested in your feedback, suggestions, clarifications, corrections and so forth. After all, one of *your* suggestions may make it into the second edition (if there is such)! Please, *write* or *send email.* Do *not* call, and please *do not* use of Springer's FAX numbers. Please do not contact me at my current employer.

Here is the address at which I can be contacted:

Stephan von Bechtolsheim
Springer-Verlag
175 Fifth Avenue
New York, NY 10010
Email: `texip@cs.purdue.edu`

The above email address will either forward to me or to someone at Springer-Verlag who will subsequently get in touch with me.

7 TEX Users Group

TEX is a great product and it's free! I do encourage you to show your support by joining the TEX Users Group, which can be contacted at the following address:

TEX Users Group
P.O. Box 869
Santa Barbara, CA 93102
(805) 899 4673
Internet: `tug@math.ams.com`

Corrections to this series will be distributed by the TeX Users Group. This is yet another reason to join this organization.

8 Rewriting Code for Improved Readability

Let me point out that I took the liberty of reformatting plain format macro code for improved readability. For instance, to set a register (counter or dimension register) to zero, the plain format declares a dimension register `\@z` which is initialized to zero:

```
\newdimen\z@
\z@ = 0pt
```

This register can be used to set another register, let's say `\dimen0`, to zero, as in `\dimen0 = \z@`. This, in turn, can be abbreviated to `\dimen0 \z@`. The *dimension* register `\z@` can also be used to set a *counter* registers to zero, as in `\count0 = \z@`. Finally, `\z@` can be used in place of a number 0, that is you can write `\count3 = \z@` or `\count3 \z@`, for short. This method of using `\z@` has also the advantage that no space needs to be written after the constant.

Now back to the changes to the plain format macros I applied in this series, when I reprint any plain format macro or macros. For instance, `\z@` is used in the definition of `\allowbreak` as follows:

```
\def\allowbreak{\penalty\z@}
```

In this series though, I state that this macro is defined as follows:

```
\def\allowbreak{\penalty 0 }
```

I think the second form of presenting this macro is more *readable*, and this is really all that this is about. I prefer readability and clarity in computer code over efficiency.

9 Summary

I hope you have a good time reading this series.

10 Building Paragraphs in TeX

This and the following two chapters deal with typesetting paragraphs. Paragraphs are a very important and basic entity in TeX. With the help of TeX, rather complicated paragraph shapes can be generated—let's start with simple ones though.

10.1 The Basics

This section explains the basics of the processing of a paragraph by TeX. You will learn how to start and how to end a paragraph and what processing TeX does in those cases.

Starting and ending a paragraph can be summarized as follows: to *start* a paragraph usually no special instruction is required; just enter your text. To *end* a paragraph, simply leave an empty line after the paragraph's text.

10.1.1 Spaces, Returns, etc.

Let me briefly exemplify some of the rules mentioned in 2.8, p. I-23, on the treatment of spaces and returns in the typesetting of ordinary text. For that purpose, I present two examples of identical text. In the first example, the text is entered the usual way: each input line is around 60 characters long. In the second example, changes in the input are made.

Here is the example. First input A:

```
$$
    \vbox{
        \hsize = 26pc
        \hrule
        \medskip

        This is just a little paragraph. We type some text so we
```

```
        can show the reader some of the rules about spaces and
        returns. It is important to understand those rules. The
        reader may consider experimenting on his or her own
        along the lines described here.

        \medskip
        \hrule
    }
$$
```

Next input B:

```
$$
    \vbox{
        \hsize = 26pc
        \hrule
        \medskip

            This     is just a little        paragraph.
        We type some
        text
        so we can show the reader some of the rules about
                sp%
                a%
                    ce%
                s
                   and
        returns. It is          important
        to understand those rules. The reader may consider
        experimenting on
                        his or her
            own
                           along the lines described here.

        \medskip
        \hrule
    }
$$
```

Both inputs generate the following identical output:

This is just a little paragraph. We type some text so we can show the reader some of the rules about spaces and returns. It is important to understand those rules. The reader may consider experimenting on his or her own along the lines described here.

Input A (the way you would probably enter the preceding paragraph) and input B (a way you probably would never enter a paragraph) differ as far as the following items are concerned. The purpose of the following list is to summarize those factors which are ignored by TEX when it comes to line break computation when a paragraph is put together.

1. Different line breaks are used in the second input. There is no change in the end result because returns and spaces are treated identically by TEX; see 2.8, item 5, p. I-25.
2. Multiple spaces instead of simple spaces are inserted between some of the words in the second input. Again there is no change in the output because adjacent multiple spaces are reduced to simple spaces by TEX; see 2.8, item 2, p. I-24.
3. Some of the lines of the second input are indented in an arbitrary fashion. There will again be no change in the printed output because spaces at the beginning of a line are ignored; see 2.8, item 9, p. I-26.
4. Some of the words in the second input are split up across multiple lines. Each split point is immediately followed by a "%" to suppress the space normally resulting from the return character. There is no change in the final output because a percent sign makes TEX ignore the end of line character, and the spaces at the beginning of the next line are ignored anyway; see 2.8, item 9, p. I-27.

10.1.2 The Basics of Processing a Paragraph

TEX handles a paragraph as follows (it is oversimplified, but sufficient at this point).

1. TEX *first reads the* **whole** *text of the paragraph.* All returns become spaces (more accurately, space tokens) and thus the whole paragraph becomes one long line. No line break decisions are made *while* the paragraph is read-in.
2. TEX *finds the line breaks.* The single line formed in the preceding step is now divided into lines *approximately* `\hsize` wide. `\hsize` is the horizontal size, the width of a line of a paragraph. The actual algorithm is *far more complicated and sophisticated.* See 12.7, p. 137, for details.
3. TEX *inserts each line into an* `\hbox to \hsize`. There are three possibilities:
   (a) The line is *too short.* The interword glue is stretched so that the line becomes as wide as specified by `\hsize`.
   (b) The line has *exactly* the right length (this case is rather unlikely, but definitely possible). The interword glue does not need to be adjusted.
   (c) The line is *too long.* The interword glue is shrunk so that the line becomes as wide as specified by `\hsize`.

10.1.3 The Operation of the Line Breaking Algorithm

The idea of breaking a paragraph into lines of approximately the same length and then adjusting the length of each line is demonstrated in the following example. For simplicity assume the first line of the paragraph is not indented.

I use 3.04 in for the value of `\hsize` in this example. This value does not cause TeX to generate warning messages about underfull or overfull lines, which indicate that TeX was not able to find satisfactory line breaks. Since compound words (there are quite a number of these in the following example text) are not hyphened by TeX, it is rather difficult for TeX to typeset this paragraph without excessively stretching and shrinking lines.

The following experiment is performed in three steps:

1. A paragraph is entered, processed by TeX, and printed.
2. An hbox is generated *of every line of the printed version of the paragraph.* This shows the lines *before* they are pushed together or stretched to adjust their length, and achieve left and right justified text. Therefore, the distances between all words are precisely the same (ignoring the increased space between a period ending a sentence and the following word). The text now is *not* right justified but ragged right.
3. Now each `\hbox` command of the preceding step is changed to an `\hbox to \hsize` (or `\line`) command. Now the original paragraph output is printed again and the experiment, which simulates the operation of the line breaking algorithm in TeX, is completed.

In a certain sense, this experiment is "reverse engineering": first we let TeX do the work (step 1 of the above list), then we look at the output before the line lengths are adjusted (step 2), and then (step 3) the line lengths of all the lines are adjusted again.

Step 1. Here is the input for the paragraph (the text is from Shakespeare (1605)). Display math mode is used to center the paragraph and the paragraph is enclosed inside a vbox; see 10.3, p. 8. The output of the following input is reprinted in the top third of Figure 10.1, p. 6.

• `ex-pp1.tip` •

```
$$
    \vbox{
        \hsize = 3.04 in
        \noindent A knave, a rascal, an eater of broken meats; a
        base, proud, shallow, beggarly, three-suited, hundred-pound,
        filthy worsted-stocking knave; a lily-livered,
        actiontaking, whoreson, glass-gazing, superserviceable,
        finical rogue; one-trunk-inheriting slave; one that wouldst
        be a bawd in way of good service, and art nothing but the
        composition of a knave, beggar, cowards, pander, and the
        son and heir of a mongrel bitch; one whom I will beat
        into clamorous whining if thou deny'st the least syllable
        of thy addition.
    }
$$
```

• End of `ex-pp1.tip` •

Step 2. The following TeX input regenerates the paragraph without the adjustment of the line length of each line. In other words, lines that are wider than `\hsize` have not yet been pushed together (by shrinking the interword glue) to the width of `\hsize`. Similarly, lines that are too short (shorter than `\hsize`) have not yet been spread-out to the full length (again to the width of `\hsize`, this time by stretching the interword glue). The output generated by the following source code is reprinted in the middle of Figure 10.1 on the next page.

• `ex-pp2.tip` •

```
$$
    \vbox{
        \hsize = 3.04in
        \hbox{A knave, a rascal, an eater of broken meats; a base,}
        \hbox{proud, shallow, beggarly, three-suited, hundred-}
        \hbox{pound, filthy worsted-stocking knave; a lily-livered,}
        \hbox{actiontaking, whoreson, glass-gazing, superservice-}
        \hbox{able, finical rogue; one-trunk-inheriting slave; one}
        \hbox{that wouldst be a bawd in way of good service,}
        \hbox{and art nothing but the composition of a knave,}
        \hbox{beggar, cowards, pander, and the son and heir of}
        \hbox{a mongrel bitch; one whom I will beat into clam-}
        \hbox{orous whining if thou deny'st the least syllable of}
        \hbox{thy addition.}
    }
$$
```

• End of `ex-pp2.tip` •

Step 3. Now replace each `\hbox` with an `\hbox to \hsize` (which is equivalent to `\line`; see 6.6, p. I-185). This causes the original paragraph to be printed, because the interword spaces are stretched and pushed together. Here is the new input. The new output, reprinted in the bottom third of Figure 10.1 on the next page is identical to the output generated in step 1 (top third of the same figure).

• `ex-pp3.tip` •

```
$$
    \vbox{
        \hsize = 3.04 in
        \line{A knave, a rascal, an eater of broken meats; a base,}
        \line{proud, shallow, beggarly, three-suited, hundred-}
        \line{pound, filthy worsted-stocking knave; a lily-livered,}
        \line{actiontaking, whoreson, glass-gazing, superservice-}
        \line{able, finical rogue; one-trunk-inheriting slave; one}
        \line{that wouldst be a bawd in way of good service,}
        \line{and art nothing but the composition of a knave,}
        \line{beggar, cowards, pander, and the son and heir of}
        \line{a mongrel bitch; one whom I will beat into clam-}
        \line{orous whining if thou deny'st the least syllable of}
        \hbox{thy addition.}    % last line: \hbox, not \line.
    }
$$
```

• End of `ex-pp3.tip` •

The following output is generated by the original paragraph (step 1):

> A knave, a rascal, an eater of broken meats; a base, proud, shallow, beggarly, three-suited, hundred-pound, filthy worsted-stocking knave; a lily-livered, actiontaking, whoreson, glass-gazing, superserviceable, finical rogue; one-trunk-inheriting slave; one that wouldst be a bawd in way of good service, and art nothing but the composition of a knave, beggar, cowards, pander, and the son and heir of a mongrel bitch; one whom I will beat into clamorous whining if thou deny'st the least syllable of thy addition.

The following output is generated by inserting every line into an hbox (step 2):

> A knave, a rascal, an eater of broken meats; a base,
> proud, shallow, beggarly, three-suited, hundred-
> pound, filthy worsted-stocking knave; a lily-livered,
> actiontaking, whoreson, glass-gazing, superservice-
> able, finical rogue; one-trunk-inheriting slave; one
> that wouldst be a bawd in way of good service,
> and art nothing but the composition of a knave,
> beggar, cowards, pander, and the son and heir of
> a mongrel bitch; one whom I will beat into clam-
> orous whining if thou deny'st the least syllable of
> thy addition.

The following output is generated by inserting every line into an `\hbox to \hsize` (step 3):

> A knave, a rascal, an eater of broken meats; a base,
> proud, shallow, beggarly, three-suited, hundred-
> pound, filthy worsted-stocking knave; a lily-livered,
> actiontaking, whoreson, glass-gazing, superservice-
> able, finical rogue; one-trunk-inheriting slave; one
> that wouldst be a bawd in way of good service,
> and art nothing but the composition of a knave,
> beggar, cowards, pander, and the son and heir of
> a mongrel bitch; one whom I will beat into clam-
> orous whining if thou deny'st the least syllable of
> thy addition.

Figure 10.1. Paragraph typesetting example.

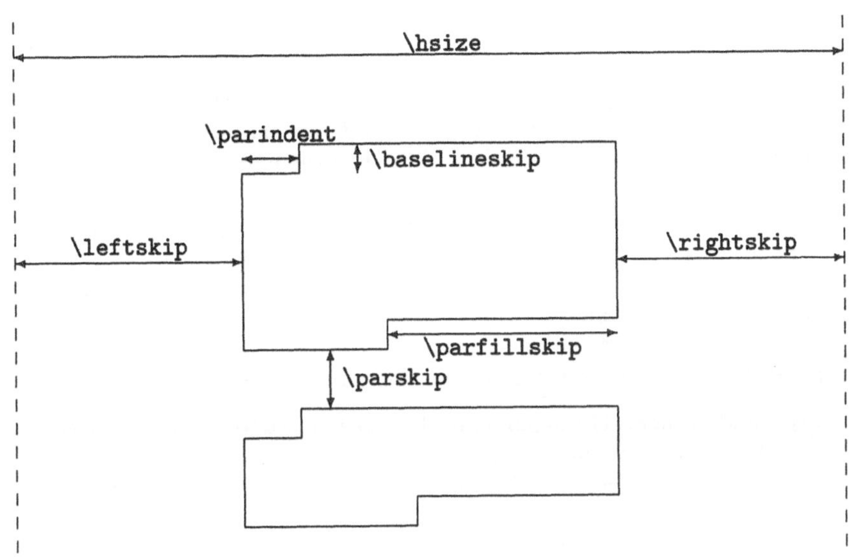

Figure 10.2. The basic parameters of paragraph layout.

10.2 Parameters Defining Paragraph Shapes

The following is a brief overview of the most important parameters that control the layout of paragraphs. Some of the parameters are included in Fig. 10.2 on this page. All the parameters listed here are explained in greater detail in the remainder of this chapter and the next.

1. The *line spacing* of a paragraph is controlled by `\baselineskip`, `\lineskip` and `\lineskiplimit`. See 7.3.4, p. I-220, for details.
2. The *paragraph indentation* is controlled by `\parindent`. See 10.7, p. 18, for details.
3. The *vertical space between paragraphs* is controlled by `\parskip`. See 10.8, p. 25, for details.
4. The line width is determined by the value stored in the dimension parameter `\hsize` (horizontal size). The line width is referred to as the line length.

   `\hsize` does *not* include the white marginal space to the left and the right of paragraphs and other output. The default of the plain format is 6.5 in. This leaves 1 in of white margin to the left and right of the paragraphs for a paper width of 8.5 in. The fact that the 6.5 in wide "TEX page" is placed symmetrically (with 1 in wide margins on both sides) has to do with a *convention* for device drivers, which are supposed to place the origin of TEX's coordinate system 1 in to the right (and 1 in down) from the upper left corner of the paper on which the output appears.
5. The *width of a paragraph*, in addition to `\hsize`, is determined by `\leftskip`

and **\rightskip**; see 10.6, p. 14. Both parameters are usually zero, TEX's default setting.

\leftskip determines the amount by which the left margin of a paragraph is shifted to the right. **\rightskip** determines the amount by which the right margin of a paragraph is shifted to the left.

6. The *interword spacing* is normally determined by the currently active font. The user can define his own interword spacing by changing **\spaceskip** and **\xspaceskip**. See 16.2, p. 275, and 16.2.4, p. 277, for details.
7. The *length of the last line of a paragraph* is controlled by **\parfillskip**. See 10.11, p. 45, for details.
8. *Line breaking* is influenced by penalties and demerits of the line breaking algorithm. The following parameters play an important role in line breaking:

   (a) **\tolerance**. The maximum allowable tolerance if pass 2 of the line breaking algorithm is executed. See 12.7.2, p. 137, for details.
   (b) **\pretolerance**. Same for pass 1. See 12.7.2, p. 137, for details.
   (c) **\linepenalty**. See 12.7.9, p. 143, for details.
   (d) **\hyphenpenalty**. See 12.7.10, item 1, p. 144, for details.
   (e) **\exhyphenpenalty**. See 12.7.10, item 2, p. 144, for details.
   (f) **\adjdemerits**. Demerit added for every pair of two visually incompatible pairs of lines. See 12.7.15, p. 148, for details.
   (g) **\doublehyphendemerits**. Demerit added for two adjacent hyphenated lines. See 12.7.15, p. 148, for details.
   (h) **\finalhyphendemerits**. See 12.7.15, p. 148, for details.

9. *Hanging indentation* can be generated very easily in TEX by loading the parameters **\hangafter** and **\hangindent**; see 11.4, p. 80, for details.

Arbitrary paragraph shapes in TEX can be generated by using **\parshape**; see 12.1, p. 105.

10.3 Paragraphs and Vboxes

A paragraph can be enclosed inside a vertical box (this was shown before). Enclosing a paragraph inside a vbox prevents TEX from generating a page break in the middle of a paragraph. Therefore the paragraph will always appear as one unit on one page.

The **\vbox** command causes an implicit group to be established and therefore **\hsize**, and all the other parameters related to paragraph shapes, can be changed inside such a box without affecting the settings outside the box. This comes in very handy as the following example shows. The following input was used (**\VboxR** prints rules surrounding the box):

```
\InputD{box-mac.tip}                    % 9.3.14, p. I-343.
\VboxR{%
```

```
    \hsize = 24pc
    \medskip

    This is a paragraph which has been typeset by enclosing it
    inside a vbox. Observe that the paragraph cannot
    be broken across pages and that the change of {\tt\string\hsize}
    is local to this vbox.

        By the way, one can have more than one paragraph inside a
    vbox and then both paragraphs will appear on the same
    page in the output.

    \medskip
}
```

This input generated the following output:

This is a paragraph which has been typeset by enclosing it inside a vbox. Observe that the paragraph cannot be broken across pages and that the change of `\hsize` is local to this vbox.

By the way, one can have more than one paragraph inside a vbox and then both paragraphs will appear on the same page in the output.

10.3.1 Centering a Paragraph Enclosed in a Vbox

Typically a paragraph like the one just discussed is centered on the page when printed. This can be done by enclosing the paragraph inside a vbox and then enclosing the vbox inside the display math mode.

For example, the following source code,

```
$$
    \VboxR{%
        \hsize = 24pc
        This is a paragraph which has been typeset by enclosing
        it in a vbox. This vbox is centered using the
        display math mode. This works very nicely as you
        can see.
    }
$$
```

generates the following output:

This is a paragraph which has been typeset by enclosing it in a vbox. This vbox is centered using the display math mode. This works very nicely as you can see.

The use of the display math mode to center text enclosed in a vbox extends to centering tables; see 40.1, p. IV-295.

10.3.2 A Paragraph with a Predetermined First Line

Assume you want to typeset a paragraph in which you determine the width of the paragraph. And you want to specify this width *indirectly* by specifying the first line of the paragraph separately from the rest of the paragraph. Here is how this is done (it is assumed that the first line of the paragraph is not indented.)

• `ex-cpar-fl.tip` •

Usual beginning: display math mode enclosing a vbox.

```
$$
\vbox{
```

Set the box register 0 to the first line of the paragraph to measure its length.

```
    \setbox0 = \hbox{This is the first line of a very short
                    paragraph,%
                }
```

Set `\hsize` to the width of the first line.

```
    \hsize = \wd0
```

Typeset paragraph starting with the first line (not indented). The `\unhbox` command retrieves the first line (see 8.1, p. I-261, for details on `\unhbox`).

```
    \noindent
    \unhbox 0
```

Now generate a space between the first line and the second line. This space later disappears in the line break between the first and second line. The space must be present because otherwise TEX cannot find a line break between the end of the first line and the beginning of the second line.

```
    \space
```

The rest of the paragraph's text follows:

```
    which was specified by the user with the intention
    that it controls the width of the whole paragraph.
    That should be enough for the example.
}
$$
```

• End of `ex-cpar-fl.tip` •

The preceding TEX source code generated the following output:

> This is the first line of a very short paragraph, which was specified by the user with the intention that it controls the width of the whole paragraph. That should be enough for the example.

In 21.10.2, p. III-201, you find a macro implementing the ideas of the preceding example.

10.4 Starting a Paragraph

Now we will discuss starting a paragraph in TEX in more detail; there are cases where one has to have a precise understanding of when TEX starts a paragraph.

10.4.1 Ways to Start a Paragraph

Instructing TEX to start a paragraph is something you probably have not thought about because you simply entered your text and TEX took care of your paragraphs.

A list of circumstances in which TEX starts a paragraph follows. These circumstances assume that TEX is in one of its vertical modes. In other words, TEX has not yet started a paragraph.

1. A *character* (the first character of a paragraph) is the most common way for starting a paragraph. Note that this first character has a dual function:
   (a) It triggers the starting of the paragraph.
   (b) It causes TEX to typeset that particular character.
2. A *mathematical formula,* be it in inline math mode (**$**) or in display math mode (**$$**), starts a paragraph. This does not occur very frequently, but it is definitely a possibility. Here is a short example:

```
Last line of previous paragraph.

$x+1 > x$ is a mathematical equation which starts
this paragraph.
```

3. The instruction, **\leavevmode** (leave vertical mode), is an explicit instruction in TEX to start a paragraph. This macro is in most cases applied if one needs to start a paragraph with an hbox, because hboxes do not cause TEX to begin a paragraph.

   \leavevmode, when applied in the middle of a paragraph, has no effect (the same way **\par** has no effect if TEX is already in vertical mode).

   The sequence **\leavevmode\par**, when applied in vertical mode, causes TEX to start a paragraph and immediately end it. This generates vertical glue in the amount of **\parskip** (see 10.8, p. 25) + **\baselineskip**. See 18.2.5, p. III-19, for an application.
4. **\noindent** and **\indent** both start paragraphs; see 10.7.1, p. 19, and 10.7.2, p. 19.
5. *Horizontal glue* (**\hskip**) starts a paragraph. For instance, **\hskip 30pt** would indent the first line of a paragraph by 30 pt. Additionally TEX adds the regular paragraph indentation **\parindent** (see 10.7, p. 18, for details).
6. A *vertical* rule (**\vrule**) begins a paragraph; see 5.7, p. I-154.
7. **\␣** (control space) starts a paragraph.

8. `\accent` starts a paragraph.
9. `\char` starts a paragraph.

Remember that an hbox does *not* start a paragraph. If the very first item of a paragraph is supposed to be an hbox, `\leavevmode` must be used; see 7.2.3, p. I-214.

10.4.2 Computations Performed by TEX at the Beginning of a Paragraph

TEX performs certain computations at the beginning of a paragraph. (*How* TEX is instructed to start a paragraph has no influence on the following actions). However, the *order* of the following events is very important for later discussions.

1. TEX inserts `\parskip` glue, an implicit vertical glue, into the current vertical list. This glue determines the vertical spacing of the text preceding the paragraph (frequently another paragraph) and the current paragraph.

   There are a few instances where no `\parskip` glue is inserted (for instance, at the beginning of a vbox); see 7.2.1, p. I-212, for an example.

   Observe that the `\parskip` glue is inserted at the *beginning* of a paragraph, not at the end. Frequently it is sufficient to think of `\parskip` glue as being inserted between paragraphs, but sometimes one must be more accurate; see 10.8, p. 25, for details.
2. TEX sets `\prevgraf` to zero. This counter parameter, at the end of typesetting the current paragraph, contains the number of lines of the paragraph just typeset. Displayed equations occurring in the middle of a paragraph influence the value assigned to this register. See 12.5, p. 130, for details.
3. TEX switches to horizontal mode.
4. TEX adds the paragraph indentation `\parindent` to the beginning of the paragraph. The indentation of paragraphs is discussed in 10.7, p. 18.
5. TEX evaluates `\everypar`, a token parameter (see 10.9, p. 27, for details). This evaluation happens *after* TEX has entered horizontal mode, and therefore, *after* TEX has generated the `\parskip` glue and written it to the main vertical list.

10.5 Ending a Paragraph

Paragraphs are terminated in various ways. A precise understanding of the termination methods is important.

10.5.1 The \par-Primitive, \endgraf

The **\par** primitive's main purpose is to end a paragraph. We need to discuss this primitive now.

1. TeX converts empty lines into **\par** tokens. This is why empty lines can and should be used to separate paragraphs.
2. TeX can be typeset so each *return* in the input is converted into a separate line in the output—in other words, where TeX follows the line breaks of the input. One way to achieve this is to add **\par** to the end of every line of the input. See 10.10, p. 36, for a more detailed discussion.
3. **\par** tokens are *not* allowed in either inline or display math mode. Therefore, empty lines within mathematical formulas are also illegal.
4. **\par** in vertical mode is ignored. A consequence of this is that, if in doubt, you can force TeX to enter vertical mode by writing **\par**, and if TeX is already in vertical mode, the **\par** has no effect.
5. **\par** in restricted horizontal mode is ignored.
6. A **\par** in the argument of a macro is illegal, and a "runaway argument error" is generated by TeX. If a macro is declared as "long" (which is the exception rather than the rule), **\par**s in arguments are valid; see 22.6, p. III-226, for details.
7. The plain format defines **\endgraf** as equivalent to **\par** by the following instruction: **\let\endgraf = \par**. Therefore, **\endgraf** can be used to terminate a paragraph if **\par** was redefined as is done in a few instances; see 23.2.3, p. III-245.

10.5.2 Computations Performed at the End of a Paragraph

TeX does *not* typeset the paragraph until it has found the end of a paragraph. Here is a brief sketch of what happens once the end of the paragraph is found (so far the whole paragraph has only been read-in; also note that that the insertion of **\parskip** glue, the insertion of the paragraph indentation (**\parindent**), and the evaluation of **\everypar** has already taken place):

1. TeX breaks up the text into lines of approximately the same length. Any special paragraph shapes, paragraph indentations, and various other factors are ignored at that point in time. Also the actual algorithm is far more complicated.
2. TeX uses the current valid values of **\leftskip**, **\rightskip**, **\baselineskip** to typeset the paragraph.

   The important point here is that these values apply to the whole text of the paragraph. In particular, it does not matter what the values were at the beginning of the paragraph. It also does not matter how often these values were changed in the middle of a paragraph. The only thing that matters is

the value of these registers at the very end of a paragraph. See 12.2, p. 110, for details.
3. Each typeset line of the paragraph is written to the main vertical list (the main vertical list is where the material to be printed on the current page is collected; see 32.1.2, p. IV-2, for more details). Naturally, interline glue is inserted between the lines.
4. TEX continues in vertical mode.

An important observation is that there is no way for the user manually to intervene in the above process. There is not, for instance, a **\everyline** type of token register that is evaluated each time a line of a paragraph is written to the main vertical list, and there is no way to intervene between steps 2 and 3 as they are executed by TEX. If one needs to apply an operation to every line of a paragraph, it must be done differently; see 8.2.12, p. I-296.

10.6 Indenting Paragraphs, \leftskip and \rightskip

\leftskip and **\rightskip** are two glue parameters. They determine implicit horizontal glues that are inserted to the *left* and *right* of *each line of a paragraph.* The same amount of glue is inserted for all lines of one paragraph, in other words, there is only one **\leftskip** and one **\rightskip** value per paragraph.

The default values for **\leftskip** and **\rightskip** are zero. In effect, the length of a line in a paragraph is reduced to **\hsize** − **\leftskip** − **\rightskip**. See Fig. 10.2, p. 7, for a graphical demonstration of the three parameters **\hsize**, **\leftskip** and **\rightskip**.

10.6.1 Applications of \leftskip and \rightskip

An overview of the applications of **\leftskip**, and **\rightskip**, and a number of corresponding examples follow.

1. **\leftskip** and **\rightskip** can be used to generate indented paragraphs; see 10.6.2 on the next page.
2. The macro **\narrower** changes **\leftskip** and **\rightskip** to generate indented paragraphs; see 10.6.3, p. 17.
3. These two registers can be used to typeset bibliographies. See 10.7.4, p. 21, for an example.
4. Dictionary-like entries can be typeset by modifying these two parameters. See 10.7.5, p. 22, for an example.

10.6.2 Changing \leftskip and \rightskip

When **\leftskip** and **\rightskip** are assigned new values, these values stay in effect until changed again. Typically changes to these values are enclosed inside groups, and the values of these parameters are of course affected by the end of a group. The values of these registers are *not* automatically reset to zero at the beginning of each paragraph.

The most common application of **\leftskip** and **\rightskip** is to generate *indented paragraphs*. Fig. 10.2, p. 7, may be helpful in understanding this example.

In these examples, observe where empty lines are placed to finish a paragraph. A paragraph must be finished either by an empty line or a **\par** *before* changes to **\leftskip** and **\rightskip** are undone by a closing curly brace finishing a group. See 12.2, p. 110, for details.

Here is the source code of an example:

```
\medskip
\hrule
\medskip

\parskip = \baselineskip
\parindent = 20pt

    {\tt [1]} This is a paragraph for the beginning. This paragraph
is not indented on either side. So we now start two
indented paragraphs. Here we go!

{   \leftskip = 30pt
    \rightskip = 30pt
        {\tt [2]} Here we have a paragraph that is
    indented equally on both sides. This is not really
    necessary---the paragraph could be indented by different
    amounts on the left and right side.

        {\tt [3]} Here is a second paragraph. Same deal! We
    now show a paragraph that is indented even further. We
    make the indentation symmetrical again, but this is not
    really necessary.

    {   \advance\leftskip by 30pt          % \leftskip now 60pt
        \rightskip = \leftskip             % Symmetrical.
        \parindent = 0pt                   % No paragraph indentation.
            {\tt [4]} So now we have two paragraphs, which are
        indented even further. So now we have two paragraphs,
        which are indented even further. So now we have two
        paragraphs, which are indented even further.

            {\tt [5]} So now we have two paragraphs, which are
        indented even further. So now we have two paragraphs, which
        are indented even further.
```

```

    }
        {\tt [6]} This paragraph here is out on the first level
    of indentation. This paragraph here is out on the first level.
    This paragraph here is out on the first level.
    This paragraph here is out on the first level.
    This paragraph here is out on the first level.

}
    {\tt [7]} And now we are back to text on the outermost level.
And now we are back to text on the outermost level.
And now we are back to text on the outermost level.
And now we are back to text on the outermost level.

\medskip
\hrule
\medskip
```

This source code generated the following output:

---

[1] This is a paragraph for the beginning. This paragraph is not indented on either side. So we now start two indented paragraphs. Here we go!

> [2] Here we have a paragraph that is indented equally on both sides. This is not really necessary—the paragraph could be indented by different amounts on the left and right side.
>
> [3] Here is a second paragraph. Same deal! We now show a paragraph that is indented even further. We make the indentation symmetrical again, but this is not really necessary.
>
> > [4] So now we have two paragraphs, which are indented even further. So now we have two paragraphs, which are indented even further. So now we have two paragraphs, which are indented even further.
> >
> > [5] So now we have two paragraphs, which are indented even further. So now we have two paragraphs, which are indented even further.
>
> [6] This paragraph here is out on the first level of indentation. This paragraph here is out on the first level. This paragraph here is out on the first level. This paragraph here is out on the first level. This paragraph here is out on the first level.

[7] And now we are back to text on the outermost level. And now we are back to text on the outermost level. And now we are back to text on the outermost level. And now we are back to text on the outermost level.

10.6.3 The Definition of Macro \narrower

When quoting some material, one typically indents the text on both sides by the same amount. The **\narrower** macro of the plain format increments **\leftskip** and **\rightskip** by **\parindent**. If the paragraph is enclosed with **\narrower**, the total paragraph indentation is the same as if both sides of the paragraph had been indented.

Rather than assigning **\parindent** directly to **\leftskip** and **\rightskip**, the macro uses **\advance** (see 3.2.2, p. I-38) for changing **\leftskip** and **\rightskip**.

Here is the definition of **\narrower**:

```
\def\narrower{%
    \advance\leftskip  by \parindent
    \advance\rightskip by \parindent
}
```

The following example is an application of this macro. The following source code is used:

```
$$
    \vbox{
        \hsize = 27pc
        \parindent = 20pt
        {\tt [1]} This is some text preceding a quote. This text
        precedes a quote. This text is simply to precede a quote.
        A quote will follow which is preceded by this text.
        A quote follows this preceding text.

        {%
            \narrower
            {\tt [2]} This is the quote. The quote text is here.
            Here it is, the quote text. Quoting text is followed
            by some more text.

            {%
                \narrower
                    {\tt [3]} And here we have a quote inside a
                quote which we might call a double quote or a
                quotational quote.

            }
            {\tt [4]} This is the quote. The quote text is here.
            Here it is, the quote text. Quoting text is followed
            by some more text.

        }
        {\tt [5]} This is the text following a quote. Here is
        some text that follows our quote. This was an example of the
        {\tt \string\narrower} macro used to indent paragraphs.
    }
$$
```

This source code generated the following output:

[1] This is some text preceding a quote. This text precedes a quote. This text is simply to precede a quote. A quote will follow which is preceded by this text. A quote follows this preceding text.

[2] This is the quote. The quote text is here. Here it is, the quote text. Quoting text is followed by some more text.

[3] And here we have a quote inside a quote which we might call a double quote or a quotational quote.

[4] This is the quote. The quote text is here. Here it is, the quote text. Quoting text is followed by some more text.

[5] This is the text following a quote. Here is some text that follows our quote. This was an example of the `\narrower` macro used to indent paragraphs.

10.6.4 Negative `\leftskip` and `\rightskip`

Negative values can be used for `\leftskip` and `\rightskip`. Such values are used to write text outside the margins. Here is an example. The following source code is used in this example.

```
$$
    \VboxR{
        \hsize = 23pc
        \leftskip = -20pt
        \rightskip = -20pt
        \noindent This paragraph runs beyond the left and right
        margin because both {\tt\string\leftskip} and
        {\tt\string\rightskip} are set to negative values.
        You can even do that!!!
    }
$$
```

This input generates the following output:

This paragraph runs beyond the left and right margin because both `\leftskip` and `\rightskip` are set to negative values. You can even do that!!!

10.7 Paragraph Indentation, `\parindent`

The TEX dimension parameter `\parindent` specifies the *indentation of the first line of a paragraph.*

Indenting the first line of a paragraph in your *input* makes the input more readable on the terminal screen, but as far as TEX is concerned, the indentation in the source code does not matter because spaces at the beginning of a line are ignored anyway; see 2.8, item 9, p. I-26.

The default for **\parindent** in the plain format is 20 pt. This series, on the other hand, has been typeset with a paragraph indentation of 15 pt.

The value of **\parindent** is also used in the **\narrower** macro; see 10.6.3, p. 17.

10.7.1 Suppressing Paragraph Indentation, \noindent

In some styles (this series is an example), *the first paragraph following a heading* is *not* indented; all other paragraphs are indented. The **\noindent** primitive does just that, causing a paragraph preceded by **\noindent** not to be indented. The indentation of all following paragraphs is **\parindent** unless **\noindent** is used again (or **\parindent** is zero in the first place).

It should be emphasized that not indenting paragraphs after headings is a stylistic decision—there is no right or wrong way.

\noindent *must not* be followed by an empty line. In other words, the text of the paragraph (which is not to be indented) must follow **\noindent** immediately.

\noindent is an instruction that also starts a paragraph by putting TEX into horizontal mode. This is why it must not be followed by an empty line. Otherwise, TEX terminates the paragraph it just began (by **\noindent**) and the following text forms a separate and indented paragraph. **\noindent** followed by an empty line is equivalent to **\noindent\par**, which is equivalent to **\leavevmode\par**; see 18.2.5, p. III-19, for a discussion of the latter.

\noindent is used to start an unindented paragraph that has an hbox. Remember that hboxes alone do not start a paragraph; see 10.4.1, p. 11.

10.7.2 The Meaning of the Primitive \indent

The **\indent** primitive starts an indented paragraph. Because TEX indents the first line of a paragraph anyway, it is natural to ask why this instruction is necessary in the first place. Look at the following example and you will see that regardless of the use of **\indent** or not, the first line of the paragraph is indented:

```
$$
    \vbox{
        \hsize = 27pc

            This is a short paragraph. Let me simply
        enter some short text so that this paragraph
        has at least two lines.

```

```
        \indent This is a short paragraph. Let me simply
        enter some short text so that this paragraph
        has at least two lines.
    }
$$
```

The output generated by this source code reads:

> This is a short paragraph. Let me simply enter some short text so that this paragraph has at least two lines.
>
> This is a short paragraph. Let me simply enter some short text so that this paragraph has at least two lines.

The usual application of `\indent` is to start an indented paragraph that has an hbox. For an application see 11.4.7, p. 87. `\indent` can also be used to double the indentation of one paragraph as the next example shows. The following input is used:

```
$$
    \vbox{
        \hsize = 27pc
        \parindent = 20pt % Repeated for clarity.
        \indent\indent This paragraph's first line,
        preceded by two {\tt\string\indent}s, is indented
        twice the regular amount. This paragraph's first line
        is indented twice the regular amount.
        This paragraph's first line is indented twice
        the regular amount. Enough is enough.
    }
$$
```

This input generated the following output:

> This paragraph's first line, preceded by two `\indents`, is indented twice the regular amount. This paragraph's first line is indented twice the regular amount. This paragraph's first line is indented twice the regular amount. Enough is enough.

10.7.3 Negative Paragraph Indentation

It is possible to have a negative paragraph indentation. This causes the first line of a paragraph to "stick out" to the left of the margin. To move in the left margin of the paragraph, set `\leftskip` to some positive dimension. You can also move in the right margin by setting `\rightskip`. Both parameters were explained in 10.6, p. 14.

This example uses the following input:

```
{
    \leftskip = 1.0in
    \rightskip = 1.0in
```

```
    \parindent = -30pt
        Here is the example of a paragraph with negative paragraph
    indentation. The first line of the text extends to the left
    of the margin.  This is something different for a change.

        And here is a second paragraph. Same deal---comes out also
    with an indented body, although this paragraph is not quite
    as long.

}
```

This input generated the following output:

Here is the example of a paragraph with negative paragraph indentation. The first line of the text extends to the left of the margin. This is something different for a change.

And here is a second paragraph. Same deal—comes out also with an indented body, although this paragraph is not quite as long.

10.7.4 Using Negative Paragraph Indentation for Typesetting a Bibliography

The following input to typeset a bibliography was used; note that `\parindent` is directly computed from `\leftskip`.

```
{
\leftskip = 20pt
\parindent = -\leftskip
[Molieouear 86] A.~B.~Molieouear, {\it How to Cook Eggs Underwater
Considering the Current Moon and Other Planet Positions},
Rotten Egg Journal of the Moonlighting Society, 1986, 320--334

[Styxziuwkklo 85] X.~Y.~Styxziuwkklo, {\it How to Write Silly
``How to Do Anything'' Articles}, Rotten Egg Conference 1985,
Rottenhausen

}
```

This input generated the following output:

[Molieouear 86] A. B. Molieouear, *How to Cook Eggs Underwater Considering the Current Moon and Other Planet Positions*, Rotten Egg Journal of the Moonlighting Society, 1986, 320–334

[Styxziuwkklo 85] X. Y. Styxziuwkklo, *How to Write Silly "How to Do Anything" Articles*, Rotten Egg Conference 1985, Rottenhausen

10.7.5 Using Negative Paragraph Indentation for Typesetting Dictionaries

For a dictionary, you may want to choose a rather small negative paragraph indentation such that all the keywords (which are printed in boldface) stick out a little bit. Here is an example. The following input was used:

```
$$
    \vbox{
        \hsize = 2.5in
        \leftskip = 7pt
        \parindent = -\leftskip
        {\bf Fiasko} {\it n\/} failure; {\it Am parl\/} landslide;
            \~{} {\it machen\/} to fail

        {\bf Fibel} {\it f\/} primer, spelling-book; (Spange) brooch,
            clasp.

        {\bf Fiber} {\it f\/} fibre ({\it od\/} fiber); fiberboard;

        {\bf Fichte} {\it f\/} pine(-tree); spruce;
    }
$$
```

This input generated the following output:

Fiasko *n* failure; *Am parl* landslide;
  machen to fail
Fibel *f* primer, spelling-book; (Spange)
  brooch, clasp.
Fiber *f* fibre (*od* fiber); fiberboard;
Fichte *f* pine(-tree); spruce;

See 11.4, p. 80, for a solution of this problem using hanging indentation.

10.7.6 Using Negative Paragraph Indentations for Lists

A *list* consists of paragraphs indented on the left side where these paragraphs have *item labels*, which stick out to the left of a paragraph. Lists are discussed extensively in the next chapter. The purpose of the discussion here is to lay some important groundwork.

Item labels can be either left justified (for item labels consisting of text, for instance) or right justified (for item labels consisting of numbers like (1), (2), ...). Following you find an example for both justifications.

10.7.6.1 Right Justified Item Labels

Here is the TEX source code of the first example of a list with right justified labels.

```
{
    \leftskip = 2.0in
    \rightskip = 1.0in
    \parindent = -50pt

    \leavevmode
    \HboxR to -\parindent{% Same as \hbox to 50pt
        \hfil
        9.%
        \hskip 10pt % Space between right end of label and text.
    }%
    And here is the text of a paragraph. This can go done for
    ever and ever. Let's have another one.

    \leavevmode
    \HboxR to -\parindent{%
        \hfil
        10.%
        \hskip 10pt
    }%
    So here is some more text, and even more. If you like it,
    then let's go on.

}
```

This input generated the following output:

9. And here is the text of a paragraph. This can go done for ever and ever. Let's have another one.

10. So here is some more text, and even more. If you like it, then let's go on.

10.7.6.2 Using \hbox Instead Of \HboxR

Now let me show the same examples using \hbox instead of \HboxR. All the rules around the labels disappear, and everything looks exactly as expected.

9. And here is the text of a paragraph. This can go done for ever and ever. Let's have another one.

10. So here is some more text, and

even more. If you like it, then
let's go on.

10.7.6.3 Omitting \leavevmode

The **\leavevmode** instructions preceding the **\hbox** instructions generating the item labels are mandatory. If a paragraph is supposed to begin with an hbox, **\leavevmode** must be used.

Let us examine the printed output generated by the preceding example if the **\leavevmodes** are omitted. Now each **\hbox** instruction that contains one label is printed as an independent hbox, on a line by itself. The paragraph itself follows on the next line.

Using **\leavevmode** to start a paragraph with an hbox is the most frequently occurring example of **\leavevmode**. There are other ways to start a paragraph with an hbox. For instance, **\hskip 0pt** will do so. See 10.4, p. 11, for details.

9.

And here is the text of a paragraph. This
can go done for ever and ever.
Let's have another one.

10.

So here is some more text, and even more.
If you like it, then let's go on.

10.7.6.4 Left Justified Item Labels

The setup for left justified item labels is almost identical to the setup for right justified labels with the exception of how the item label containing hbox is built. Here is the input for our example.

```
{
     \leftskip = 2.0in
     \rightskip = 1.0in
     \parindent = -50pt
     \leavevmode
     \HboxR to -\parindent{%
         \bf Label%
         \hfil
     }%
     And here is the text of a paragraph. This can go on for
     ever and ever. Let's have another one.

     \leavevmode
     \HboxR to -\parindent{%
         \bf Lab%
         \hfil
     }%
```

```
    So here is some more text, and even more. If you like it,
    then let's go on.
    \par
}
```

The preceding input generates the following output:

Label And here is the text of a paragraph. This can go on for ever and ever. Let's have another one.

Lab So here is some more text, and even more. If you like it, then let's go on.

10.8 The Vertical Spacing Between Paragraphs, \parskip

The implicit vertical glue, **\parskip** is inserted at the beginning of each paragraph. It determines the distance between the text preceding the current paragraph and the current paragraph. Because in most cases this "preceding text" is another paragraph, the title of this Section reads "The Vertical Spacing Between Paragraphs, **\parskip**," rather than something like "The Vertical Glue Inserted Before Each Paragraph," which would actually be more accurate.

Remember, the **\parskip** glue is inserted *before* and *not after* a paragraph. This distinction can be important as the example in 10.8.2 on the next page shows.

10.8.1 Setting \parskip

The default for **\parskip** in the plain format is `0pt plus 1pt`, in other words, there is no additional vertical space inserted between the paragraphs (note that interline glue between the last line of a paragraph and the first line of the next paragraph is inserted by TEX, and therefore the spacing between the last line of a paragraph and the first line of the following paragraph is handled correctly). This vertical space between paragraphs can be stretched out by a small amount (1 pt), which is almost invisible to the eye.

Observe that the **\parskip** glue is an important source for TEX to adjust the page length in order to fill all pages equally, unless a ragged bottom page layout is selected. Therefore, **\parskip** glue should always have some stretchability and shrinkability (shrinkability only if its natural width is not zero). This is why **\parskip**'s default in the plain format does not read **\parskip** `= 0pt` (no stretchability or shrinkability) but **\parskip** `= 0pt plus 1pt` instead.

The issue of stretching and shrinking vertical glue to adjust the page length is discussed in 32.4, p. IV-10.

A value of `12pt plus 2pt minus 1pt` for `\parskip` is what you might choose to have one empty line between paragraphs (assuming a value of 12 pt for `\baselineskip`).

10.8.2 An Example Involving \parskip

`\parskip` is a glue, which is inserted *before*, *not after*, a paragraph. This distinction can be important. Assume you wanted to enter a centered line between two paragraphs. You would probably try to do this first:

```
$$
    \vbox{
        \parskip = 12pt
        \hsize = 27pc

            {\tt [1]} This is a paragraph. We will finish this
        paragraph and then have two centered lines following it.
        An empty line to finish this paragraph is needed before
        {\tt\string\centerline} is entered.

        \centerline{{\tt [2]} A centered line}
        \centerline{{\tt [3]} A second centered line}

            {\tt [4]} And here we start the next paragraph.
        Observe that the empty line preceeding this paragraph
        is not necessary.
    }
$$
```

The output generated by this input reads as follows:

> [1] This is a paragraph. We will finish this paragraph and then have two centered lines following it. An empty line to finish this paragraph is needed before `\centerline` is entered.
>
> [2] A centered line
>
> [3] A second centered line
>
> [4] And here we start the next paragraph. Observe that the empty line preceeding this paragraph is not necessary.

Note: TeX did *not* insert `\parskip` glue before the first paragraph because the first paragraph is the first item of the vbox. It did insert `\parskip` glue before the second paragraph as one would expect. See 7.2.1, p. I-212, for details).

The resulting layout probably is not quite what you expected. Most likely you would want to insert a glue of the amount of `\parskip` *after* the first paragraph for symmetry, but TeX does not do that automatically. Therefore, you have to add a `\vskip\parskip`. Here is the "corrected" TeX source code.

```
$$
    \vbox{
        \parskip = 12pt
        \hsize = 27pc

            {\tt [1]} This is a paragraph. We will finish this
        paragraph and then have two centered lines following it.
        We need an empty line after this paragraph before using
        {\tt\string\centerline}. So here it is:

        \vskip\parskip                    % ADDED !!
        \centerline{{\tt [2]} A centered line}
        \centerline{{\tt [3]} A second centered line}

            {\tt [4]} And here we start the next paragraph.
        Observe that the empty line preceding this paragraph
        is not necessary.
    }
$$
```

The output generated by this input reads as follows (note the symmetrical vertical spacing around the two centered lines):

[1] This is a paragraph. We will finish this paragraph and then have two centered lines following it. We need an empty line after this paragraph before using `\centerline`. So here it is:

[2] A centered line
[3] A second centered line

[4] And here we start the next paragraph. Observe that the empty line preceding this paragraph is not necessary.

10.9 The Token Parameter \everypar

The token parameter `\everypar` is evaluated before a paragraph is started (see 10.4, p. 11). When this token parameter is evaluated, TEX has already entered horizontal mode (see 20.1, p. III-117, for details on token registers).

The parameter `\everypar` has several applications.

1. If you write `\everypar = {\message{A paragraph}}`, TEX prints "`A paragraph`" into the logfile and on the terminal screen each time a paragraph is started. See 29.5.2, p. III-521, for details on `\message`.
2. `\everypar = {$\bullet$ }` prints a bullet at the beginning of each paragraph. See 10.9.1 on the next page for the complete example.
3. Here is an example where paragraphs are automatically numbered. The first paragraph is preceded by `[1]`, the second by `[2]`, and so forth.

```
\newcount\ParCount
\ParCount = 0
\everypar = {%
    \advance\ParCount by 1
    {\tt[\the\ParCount]}
}
```

4. Paragraphs with hanging indentation can be generated using **\hangindent** and **\hangafter**; see 11.4, p. 80. Note that **\hangindent** and **\hangafter** are reset at the beginning of each paragraph. So to generate a series of paragraphs with hanging indentation, the values for **\hangafter** and **\hangindent** have to be reloaded at the beginning of each paragraph. This can be done automatically using **\everypar** as the following example shows:

```
\everypar = {%
    \hangafter = -2
    \hangindent = 40pt
}
```

5. If TEX had no mechanism for the automatic generation of the indentation of the first line, **\everypar** could be used for that purpose. Enter **\everypar = {\hskip 20pt}** to insert this horizontal skip instruction before the text of the paragraph is printed.
6. This series is an example of a document where paragraphs after headings are not indented. This can be done *without* using **\noindent** as will be discussed shortly.

10.9.1 An Example of Applying \everypar

The command **\everypar** can be used to insert a • in front of each paragraph. The following input is used:

```
{
    \leftskip = 25pt
    \rightskip = 25pt
    \everypar = {$\bullet$ }
        The following instruction was inserted before this paragraph:
    \Verb+\everypar = {$\bullet$ }+. As you can see the instruction
    causes a ``$\bullet$'' to be printed before each paragraph.

        The emphasis is on {\it each paragraph\/}---a special
    instruction is {\it not\/} necessary to print the~``$\bullet$''
    before this paragraph.

        Such a bullet is printed until \Verb+\everypar+ is
    reset or the group, which contains the setting of
    \Verb+\everypar+, is terminated.

}
```

This example generates the following output:

• The following instruction was inserted before this paragraph: \everypar = {$\bullet$ }. As you can see the instruction causes a "•" to be printed before each paragraph.

• The emphasis is on *each paragraph*—a special instruction is *not* necessary to print the "•" before this paragraph.

• Such a bullet is printed until \everypar is reset or the group, which contains the setting of \everypar, is terminated.

10.9.2 Dynamic \parskip Computations

TeX can be made to compute \parskip dynamically by setting the value of \parskip *right before* a paragraph is started. Let us assume that to do the computation of \parskip, the macro \ComputeParSkip is called.

10.9.2.1 Outlining the Problem Which Is Being Solved

To understand the complexity of the problem being solved note that TeX executes the following steps (among other things) at the beginning of a paragraph, in the following order:

1. Generate \parskip glue.
2. Switch to horizontal mode.
3. Evaluate \everypar.

The preceding list shows that one cannot use \everypar to execute \ComputeParSkip, which then in turn would set the value of \parskip.

This problem can also *not* be worked around by using a zero \parskip and letting \ComputeParSkip insert the proper amount of glue. TeX enters horizontal mode before a \vskip generated glue would be inserted, causing TeX to leave horizontal mode just started. Also \vadjust cannot be used, because \vadjust{\vskip ...} inserts the glue *between* the first and second line of the paragraph, and *not before* the first line of the paragraph.

10.9.2.2 A Work-Around

Let me now present a work around this problem. The source code of the solution offered begins here.

• pars-dyn.tip •

```
\input inputd.tip
```

Load a macro source file that causes TEX to change the plain format's output routine to write the main vertical list to the log file. After this file has been loaded, the feature is enabled.

```
\InputD{showpll.tip}                       % 35.8, p. IV-120.
\ShowPlainListstrue
```

Set `\parskip` to zero. This will be the standard value in the following experiment.

```
\parskip = 0pt
```

Declare a counter that is used by `\ComputeParSkip` to set `\parskip` to a variety of different values.

```
\newcount\CIndex
\CIndex = 0
```

`\ComputeParSkip` is a macro (no parameter) that sets `\parskip` to 2 pt, 8 pt and 12 pt in a cyclical fashion. The assignment to `\CIndex` must be done globally because `\ComputeParSkip` is expanded inside a group (see further below).

```
\def\ComputeParSkip{%
```

Write a message to the log file.

```
        \wlog{\string\ComputeParSkip: "\noexpand\CIndex is \the\CIndex"}%
```

Depending on the value of `\CIndex`, `\parskip` is set and also `\CIndex` is updated.

```
        \ifcase\CIndex
            \parskip = 2pt
            \global\CIndex = 1
        \or
            \parskip = 8pt
            \global\CIndex = 2
        \or
            \parskip = 12pt
            \global\CIndex = 0
        \fi
```

At the end write the new value for `\parskip` to the log file.

```
        \wlog{\string\ComputeParSkip: "\noexpand\parskip was set
            to \the\parskip"}%
}
```

Now initialize `\everypar`.

```
\everypar = {%
```

Do everything inside a group.

```
        {%
```

Remove the paragraph indentation box and terminate the paragraph just started, which, more or less, completely undoes the paragraph just started. The only exception is that the `\parskip` glue (value: zero) was already inserted and there is no way to remove it.

```
            \setbox0 = \lastbox
```

1. This is fun. This is fun.

2. This is fun. This is fun.

3. This is fun. This is fun.
4. This is fun. This is fun.

5. This is fun. This is fun.

6. This is fun. This is fun.

Figure 10.3. Output generated by executing **pars-dyn.tip**.

```
25            \par
```

Compute a new value for **\parskip**, clear out **\everypar**, and start a *new* paragraph, this time with the new value of **\parskip** in effect.

```
26            \ComputeParSkip
27            \everypar = {}%
28            \leavevmode
```

End the group previously started. Note the following important implications of ending the group: **\everypar** is reset again to its original value, that is the source code that is currently being described (the clearing out of **\everypar** done in line 36 is lost.) Also **\parskip** it set back to its original value of zero (line 4 of this source code).

```
29      }%
```

Initializing **\everypar** ends here.

```
30  }%
```

Now execute a couple of examples.

```
31  1. This is fun. This is fun.
32
33  2. This is fun. This is fun.
34
35  3. This is fun. This is fun.
36
37  4. This is fun. This is fun.
38
39  5. This is fun. This is fun.
40
41  6. This is fun. This is fun.
42  \bye
```

• End of **pars-dyn.tip** •

The output of the preceding source code appears in Figure 10.3 on this page.

Let us now look at the log file, which is generated by the preceding source file. Note that each paragraph now begins with two **\parskips**, the first one being zero, the default of **\parskip** in the preceding example, and the second one is the value of **\parskip** as computed by **\ComputeParSkip**.

The solution offered here is very close to what we wanted to do, although it is not quite what it should be because two instances of **\parskip** precede every paragraph.

Here is the log file of this example.

• **pars-dyn.log** •

```
This is TeX, C Version 3.14 (...)
**&/usr/local/tex/lib/fmt/plain pars-dyn.tip
(pars-dyn.tip (inputd.tip
(namedef.tip
) (inputdl.tip
\@InputDStream=\write0
))
(showpll.tip
(shboxes.tip
) (op-pagec.tip
))
\CIndex=\count26
\ComputeParSkip: "\CIndex is 0"
\ComputeParSkip: "\parskip was set to 2.0pt"
\ComputeParSkip: "\CIndex is 1"
\ComputeParSkip: "\parskip was set to 8.0pt"
\ComputeParSkip: "\CIndex is 2"
\ComputeParSkip: "\parskip was set to 12.0pt"
\ComputeParSkip: "\CIndex is 0"
\ComputeParSkip: "\parskip was set to 2.0pt"
\ComputeParSkip: "\CIndex is 1"
\ComputeParSkip: "\parskip was set to 8.0pt"
\ComputeParSkip: "\CIndex is 2"
\ComputeParSkip: "\parskip was set to 12.0pt"
\pagecontents from op-pagec.tip called.
*** \@ShowPlainLists: main vertical list ***
*** Page number (\count0): 1 ***

> \box255=
\vbox(643.20255+0.0)x469.75499, glue set 531.20255fill
.\glue(\topskip) 3.05556
.\hbox(6.94444+0.0)x469.75499, glue set 333.42146fil []
.\glue(\parskip) 0.0
.\glue(\parskip) 8.0
.\glue(\baselineskip) 5.05556
.\hbox(6.94444+0.0)x469.75499, glue set 333.42146fil []
.\glue(\parskip) 0.0
.\glue(\parskip) 12.0
.\glue(\baselineskip) 5.05556
.\hbox(6.94444+0.0)x469.75499, glue set 333.42146fil []
.\glue(\parskip) 0.0
.\glue(\parskip) 2.0
.\glue(\baselineskip) 5.05556
.\hbox(6.94444+0.0)x469.75499, glue set 333.42146fil []
```

```
.\glue(\parskip) 0.0
.\glue(\parskip) 8.0
.\glue(\baselineskip) 5.05556
.\hbox(6.94444+0.0)x469.75499, glue set 333.42146fil []
.\glue(\parskip) 0.0
.\glue(\parskip) 12.0
.\glue(\baselineskip) 5.05556
.\hbox(6.94444+0.0)x469.75499, glue set 333.42146fil []
.\glue 0.0 plus 1.0fill

<to be read again>
}
\@ShowPlainLists ...hOne {255}
\ifvoid \footins \wlog {\st...

\pagecontents ...owPlainLists}
\ifvoid \topins \wlog {\str...

\pagebody ...th \pagecontents
}
\plainoutput ...ine \pagebody
\makefootline }\advancepage...
<output> {\plainoutput
}
...
1.42 \bye

\@ShowPlainLists: no footnotes.
\@ShowPlainLists: no topinserts.
*** \@ShowPlainLists: end dump of page: 1 ***
\pagecontents: no topinserts.
[1] )
Output written on pars-dyn.dvi (1 page, 456 bytes).
```

Note that in the log file, no **\parskip** glue is shown at the very top of the page; see 7.2.1, p. I-212, for details.

10.9.3 Multiple \everypars

In writing this series I wished TEX would have a set of **\everypar** registers, because there were a variety of different functions that needed to be executed at the beginning of each paragraph. Because TEX does not have multiple **\everypars**, I initialize **\everypar** to call a series of different macros to achieve the same effect.

The source code to do this follows now. The following source code also discusses the meanings of the various macros invoked by **\everypar**. Note that when you preload the following macros you need to be careful with your own assignments to **\everypar**.

$\mathcal{P}'$ • **everypar.tip** •

Set up **\everypar** to evaluate some macros and then to reset some of these macro definitions.

```
\everypar = {%
     \EvalEveryPars
     \ClearEveryPars
}
```

Define macro **\EvalEveryParsCE** which evaluates **\EveryParC** through **\EveryParE** (what the **\EveryParX** type of macros are good for is explained later).

```
\def\EvalEveryParsCE{%
     \EveryParC
     \EveryParD
     \EveryParE
}
```

Define macro **\EvalEveryPars** which evaluates **\EveryParA** through **\EveryParE** and **\EveryParZ**.

```
\def\EvalEveryPars{%
     \EveryParA
     \EveryParB
     \EvalEveryParsCE
     \EveryParZ
}
```

Next define macro **\ClearEveryPars** which resets to empty the definitions of **\EveryParA** through **\EveryParE** but not of **\EveryParZ**.

```
\def\ClearEveryPars{%
     \gdef\EveryParA{}%
     \gdef\EveryParB{}%
     \gdef\EveryParC{}%
     \gdef\EveryParD{}%
     \gdef\EveryParE{}%
}
```

Now define the macro **\ClearEveryParsAll** which clears out the definitions of *all* **\EveryParX** type of macros, including **\EveryParZ**.

```
\def\ClearEveryParsAll{%
     \ClearEveryPars
     \gdef\EveryParZ{}%
}
```

Just to be sure: let's clear out all those macro definitions.

```
\ClearEveryParsAll
```

• End of **everypar.tip** •

Next we need to discuss the purpose of **\EveryParA**, **\EveryParB**, and so forth. Note that none of these macros have parameters. Also note that **\EveryParA** through **\EveryParE** belong to one group (after evaluation they are immediately cleared out), while the definition of **\EveryParZ** is *not* reset after the macro is evaluated.

1. **\EveryParA**. This macro is used by macro, **\SuppressNextParagraphIndentation**, to suppress the paragraph indentation of the next paragraph (to subsequently reset the indentation to its original value. See 10.9.4 on this page for details.
2. **\EveryParB**. This macro is used for the generation of marks, so the execution of **\mark** is delayed until TEX enters horizontal mode; see 31.2.9, p. III-604.
3. **\EveryParC**. This macro is used for **\writes** associated with labels; see 30.8.7, p. III-575, for details.
4. **\EveryParD**. Spare macro, currently not used. Can be used by the user.
5. **\EveryParE**. Another spare macro, currently not used. Can be used by the user.
6. **\EveryParZ** is a macro used by the display verbatim mode of 18.3.7.2, p. III-39. Note that this macro is *not* reset through the execution of **\everypar**, but through the display verbatim mode.

10.9.4 Suppressing the Paragraph Indentation Not Using \noindent

Let me now discuss how the indentation of a paragraph can be suppressed *not* using **\noindent**. The disadvantage of using **\noindent**, is that you cannot have the **\noindent** followed by an empty line or **\par** because **\noindent\par** constitutes an empty paragraph, resulting in an empty line to be printed (plus **\parskip** to be inserted before that). The following paragraph would be indented then. There are applications, though, where one would like to say something like "next time a paragraph is started, suppress its indentation" rather than "now start a paragraph with no indentation."

The following source code solves all of our problems.

$\mathcal{P}'$ • parin.tip •

```
\InputD{everypar.tip}                    % 10.9.3, p. 34.
```

The following dimension register is loaded with the default paragraph indentation of a document.

```
\newdimen\NormalParIndent
```

Now define the macro **\SetParIndent** which sets the paragraph indentation and at the same time saves this value in dimension register **\NormalParIndent**. For the following macro to work properly, you must call **\SetParIndent** in the very beginning of the processing of a document. This macro has one parameter, **#1**, the default value for the paragraph indentation.

```
\def\SetParIndent #1{%
     \NormalParIndent = #1%
     \parindent = #1%
```

```
}
```

Now define the macro `\SuppressNextParIndent` (no parameters). This macro simply sets `\parindent` to zero, and sets up `\everypar` so it will reset the paragraph indentation to its usual value. This works because of the order in which things happen: *first* the paragraph indentation is generated and *afterwards*, `\everypar` is evaluated.

```
\def\SuppressNextParIndent{%
    \global\parindent = 0pt
    \gdef\EveryParA{%
        \global\parindent = \NormalParIndent
        % \hskip-\parindent
    }%
}
```

Next define macro `\CancelSuppressNextParIndent` to suppress a preceding call of `\SuppressNextParIndent`.

```
\def\CancelSuppressNextParIndent{%
    \global\parindent = \NormalParIndent
    \gdef\EveryParA{}%
}
```

• End of `parin.tip` •

10.10 User-Controlled Line Breaks

User-controlled line breaks can be generated in TEX. For instance, to typeset some poetry or a theater play script, as is done in the following example, you must be able to control the line breaks. Three different solutions are presented:

1. Force line breaks in a paragraph with `\hfil\break`. See 10.10.1 on this page.
2. Treat each line as its own paragraph. This leads to the definition of the `\obeylines` macro. See 10.10.2, p. 38.
3. Use hboxes. Every line becomes a single hbox and within each hbox, horizontal glue is used to move text to its proper horizontal position. This is discussed in 10.10.3, p. 40.

10.10.1 Forcing Line Breaks Using `\hfil\break`

A line break in the middle of a paragraph is forced by `\hfil\break`. The `\hfil` pushes the current line's content to the left and `\break` forces the line break. Observe that `\break` is simply an abbreviation for `\penalty -10000`. The penalty in this case is a "horizontal" penalty influencing the line breaking algorithm.

Since it is shorter to write \\ instead of **\hfil\break**, we define macro \\ to expand to **\hfil\break**. Each continuous speech by one character is treated as one paragraph.

Here is an example. The following input was used (again from Shakespeare (1605)).

```
\parskip = 3pt plus 1pt minus 1pt
\leftskip = 1.0in
\parindent = -18pt

LEAR\hfil\break
    To thee and thine hereditary ever\hfil\break
    Remain this ample third of our fair kingdom\hfil\break
    No less in space, validity, and pleasure\hfil\break
    Than that conferred on Gonril.---Now, our joy,\hfil\break
    Although our last and least; to whose young love\hfil\break
    The vines of France and milk of Burgundy\hfil\break
    Strive to be interest; what can you say to draw\hfil\break
    A third more opulent than your sisters? Speak.

CORDELIA\hfil\break
    Nothing, my lord.

LEAR Nothing?

CORDELIA Nothing.

% Make life easier: define \\ for \hfil\break.
\def\\{\hfil\break}
LEAR\\
    Nothing will come of nothing. Speak again.

CORDELIA\\
    Unhappy that I am, I cannot heave\\
    My heart into my mouth. I love your Majesty\\
    According to my bond, no more nor less.

LEAR\\
    How, how, Cordelia? Mend your speech a little\\
    Lest you may mar your fortunes.
```

This input generated the following output:

LEAR
To thee and thine hereditary ever
Remain this ample third of our fair kingdom
No less in space, validity, and pleasure
Than that conferred on Gonril.—Now, our joy,
Although our last and least; to whose young love
The vines of France and milk of Burgundy
Strive to be interest; what can you say to draw
A third more opulent than your sisters? Speak.

CORDELIA
    Nothing, my lord.

LEAR Nothing?

CORDELIA Nothing.

LEAR
    Nothing will come of nothing. Speak again.

CORDELIA
    Unhappy that I am, I cannot heave
    My heart into my mouth. I love your Majesty
    According to my bond, no more nor less.

LEAR
    How, how, Cordelia? Mend your speech a little
    Lest you may mar your fortunes.

10.10.2 Every Line Becomes a Single Paragraph

Another way to print single lines is to make every line in the input a separate paragraph. That is, each paragraph is one line long. I use **`\noindent`** to typeset the character's name since none of the short paragraph speeches are preceded by **`\noindent`** they are indented, particularly with respect to the character names.

There are three different ways to make every line a short one-line paragraph in TEX:

1. Leave an empty line between the lines of the text. The vertical spacing of the source code, when looked at on a terminal screen, will be quite drastic though. The next solution is slightly better.
2. Add **`\par`** to the end of every line. But we can do better; see the next proposal.
3. Use **`\obeylines`**. This macro makes the return character equivalent to **`\par`**; see 18.2.4, p. III-18. In other words, after this macro has been invoked no special precautions are necessary.

Here is the source code of example.

```
\parskip = 0pt
\leftskip = 1.0in
\parindent = 18pt
\advance\leftskip by -\parindent

\noindent CORDELIA\hskip 1.0in Good my lord,

    You have begot me, bread me, loved me. I

    Return those duties back as are right fit,

```

```
    Obey you, love you, and most honor you.

    Why have my sisters husbands if the say

    They love you all? Haply, when I shall wed,

    That lord whose hand must take my plight shall carry

    Half my love with him, half my care and duty.

    Sure I shall never marry like my sisters,

    [To love my father all]
\smallskip
\noindent LEAR\par
    But goes thy hart with this?\par
\smallskip
\noindent CORDELIA \hskip 1.3in Ay, my good lord.\par
\smallskip
\noindent LEAR\par
    So young, and so untender?\par
\smallskip
\noindent CORDELIA\par
    So young, my lord, and true.
\smallskip

% Now use \obeylines, which makes every <return> act a \par.
\obeylines
\noindent LEAR
    Let it be so, thy truth then be thy dower
    For, by the sacred radiance of the sun,
    The mysteries of Hecate and the night,
    By all the operation of the orbs
    From whow we do exist and cease to be,
    Here I disclaim all my paternal care,
    Propinquity and property of blood,
    And as a stranger to my heart and me
    Hold thee from this for ever. The barbarous Scythian,
    Or he that makes his generation messes
    To gorge his appetite, shall to my bosom
    Be as well neighbored, pitied, and relieved,
    As thou my sometime daughter.
\noindent KENT \hskip 1.7in Good my liege---
\noindent LEAR
    Peace, Kent!
    $\ldots$
```

This input generates the following output:

CORDELIA Good my lord,
You have begot me, bread me, loved me. I
Return those duties back as are right fit,
Obey you, love you, and most honor you.

Why have my sisters husbands if the say
They love you all? Haply, when I shall wed,
That lord whose hand must take my plight shall carry
Half my love with him, half my care and duty.
Sure I shall never marry like my sisters,
[To love my father all]

LEAR
But goes thy hart with this?

CORDELIA Ay, my good lord.

LEAR
So young, and so untender?

CORDELIA
So young, my lord, and true.

LEAR
Let it be so, thy truth then be thy dower
For, by the sacred radiance of the sun,
The mysteries of Hecate and the night,
By all the operation of the orbs
From whow we do exist and cease to be,
Here I disclaim all my paternal care,
Propinquity and property of blood,
And as a stranger to my heart and me
Hold thee from this for ever. The barbarous Scythian,
Or he that makes his generation messes
To gorge his appetite, shall to my bosom
Be as well neighbored, pitied, and relieved,
As thou my sometime daughter.

KENT Good my liege—

LEAR
Peace, Kent!
...

10.10.3 Using Hboxes

Now let me show a completely different approach in which hboxes are used, one hbox per output line. I define two macros `\l` and `\ll`. Each macro generates one hbox and each hbox starts with an `\hskip` of either 1.0 in or 1.2 in.

Here is the source code of the example.

• **ex-slhb.tip** •

Both macros have one parameter, **#1**, the text of the hbox.

```
\def\l  #1{\hbox{\hskip 1.0in #1}}
\def\ll #1{\hbox{\hskip 1.2in #1}}
```

The example itself begins here.

```
\l{LEAR}
    \ll{Howl, Howl, Howl! O, you are men of stones.}
    \ll{Had I your tongues and eyes, I'ld use them so}
    \ll{That heaven's vault should crack. She's gone for ever.}
    \ll{She's dead as earth. Lend me looking glass.}
    \ll{If that her breath will mist or stain the stone,}
    \ll{Why then she lives.}
\l{KENT\hskip 1.0in Is this the promised end?}
\l{EDGAR}
    \ll{Image of that horror?}
```

• End of `ex-slhb.tip` •

The preceding source code generates the following output:

LEAR
    Howl, Howl, Howl! O, you are men of stones.
    Had I your tongues and eyes, I'ld use them so
    That heaven's vault should crack. She's gone for ever.
    She's dead as earth. Lend me looking glass.
    If that her breath will mist or stain the stone,
    Why then she lives.
KENT   Is this the promised end?
EDGAR
    Image of that horror?

The preceding macros `\l` and `\ll` can be redefined so no curly braces enclosing the text of each line need to be used. Then, the input reads:

```
\lnew LEAR
    \llnew Howl, Howl, Howl! O, you are men of stones.
    \llnew Had I your tongues and eyes, I'ld use them so
    ...
```

21.9.6, p. III-191, explains how this can be done.

10.10.4 Writing Single Lines Right Justified

Let me briefly discuss how you can print single *right justified* lines in TEX. For example, the following input:

```
{
    \parindent = 0pt
    \leftskip = 0pt plus 1fil
    \rightskip = 1.0in
    \parfillskip = 0pt
    This is an example,\break
    where every line comes out right justified,\break
    and the user determines the line breaks.

}
```

generates the following output:

This is an example,
where every line comes out right justified,
and the user determines the line breaks.

The infinite stretchability of **\leftskip** pushes the text over to the right as far as possible. **\rightskip** controls the distance of the right end of each line to the ordinary text margin. **\parfillskip**, the glue that normally allows the last line to be shorter than all the other lines, is set to zero so that the last line is pushed all the way to the right; see 10.11, p. 45.

10.10.5 Horizontal Glues in User-Determined Line Breaks

The previous discussions have shown that by using **\hfil\break**, a line break can be forced in the middle of a paragraph. (Note that this does *not* generate two separate paragraphs). Let me continue that discussion by using the following example:

```
$$
    \vbox{
        \hsize = 25pc
        \parindent = 0pt

        Line 1\hfil\break
        Line 2\hfil\break
        \hskip 1.0in Line 3\hfil\break
        \hskip 1.0in \hskip 1.0in Line 4\hfil\break
        \hbox{}\hskip 1.0in Line 5\hfil\break
        \hbox to 1.0in{}Line 6\hfil\break
        Line 7 line 7 line 7\break
        \hfil Line 8 line 8 line 8 line 8 line 8\break
        \hbox{}\hfil Line 9 line 9 line 9\break
    }
$$
```

This input generated the following output:

Line 1
Line 2
Line 3
Line 4
Line 5
Line 6
Line 7 line 7 line 7
Line 8 line 8 line 8 line 8 line 8
Line 9 line 9 line 9

Let me analyze this example.

1. `\hfil\break` generates an explicit line break in the middle of a paragraph (Line 1 and Line 2). The paragraph is *not* terminated and continues with the next line.
2. Observe that the glue immediately *following* a `\break` is ignored (Line 3 and Line 4). This also includes `\hfil` (Line 3, Line 4). Not only is a single glue item ignored, but multiple glues are ignored if they are consecutive and immediately follow a penalty such as one generated by `\break`.

   To prevent TEX from ignoring glue after a break point, one can use one of the following two methods after the `\break`:

   (a) Insert an empty hbox (`\hbox{}`). None of the following glue items follow the `\break` immediately now and they are no longer ignored (Line 5, Line 9).

   (b) Insert an empty hbox of the desired width (`\hbox to ...{}`, Line 6) instead of an `\hskip`.
3. A `\break` not preceded by an `\hfil` forces a line break, but the line in which the break occurs is spread over the full width of the line. This causes an underfull hbox to be reported (Line 7, Line 8) and prints output you probably did not intend to generate.
4. And then there is the possibility of generating a line pushed all the way to the right. See Line 9.

10.10.6 A Macro for Explicit Line Breaks in the Middle of a Paragraph, \\

As mentioned briefly in the previous Subsection, we can now define a macro \\ to force a line break in the middle of a paragraph. This macro works just like `\hfil\break`. A call to this macro can be followed by an *optional argument* enclosed in square brackets. The argument is expected to be a dimension, which specifies the amount of the indentation of the line *following* the line break. If no argument is provided (the term optional suggests that it can), a line break is generated where the following line starts left flush with all other lines of paragraphs. For example, while \\ generates a simple line break, `\\[1.0in]` generates a line break and also indent the next line by 1.0 in.

Here is the source code containing the definition of the \\ macro.

$\mathcal{P}'$ • **lbpar.tip** •

```
\InputD{futlet.tip}                          % 23.4.3, p. III-256.
\catcode`\@ = 11
```

Define macro \\ to force a line break in a paragraph. The line following this line break is either not indented (default) or is indented by the amount given by the optional argument to this macro (enclosed in square brackets, as in `\\[0.5in]`).

```
\def\\{%
```

Force a line break.

```
        \hfil\break
```

Insert an empty hbox so that any glue following later will not be ignored.

```
        \hbox{}%
```

If there is a "[" when the macro is called, execute `\@HfilBreakHskip`, otherwise execute `\ignorespaces`. Here testing for the optional argument and calling the proper macro or primitive takes place.

```
        \DoFutureLet{\ifx}{[}{\@HfilBreakHskip}{\ignorespaces}%
}
```

`\ignorespaces`, which is used in the following two macro definitions, is explained in 21.11, p. III-203.

```
\def\@HfilBreakHskip [#1]{%
        \hskip #1%
        \ignorespaces
}
\catcode`\@ = 12
```

• End of `lbpar.tip` •

Let me show an application of this macro. The following input is used:

```
\input inputd.tip
\InputD{lbpar.tip}                              % 10.10.6, p. 43.
$$
    \vbox{
        \hsize = 27pc
        \parskip = 12pt

        {\tt [1]} This is a sample of our new {\tt \string\\} macro.
        Let us see whether this all works now. The macro is not
        very difficult, so not much text is needed.
        Here we force a line break.\\
        {\tt [2]} And here is another one.\\
        {\tt [3]} And now let us force a line break where the
        following line is a little bit indented as it is
        here.\\[\parindent]
            {\tt [4]} So this is now indented.\\[2\parindent]
                {\tt [5]} This is now indented by twice
                        the amount of the paragraph
                        indentation.\\[\parindent]
            {\tt [6]} This line is again indented by
                            {\tt\string\parindent}.\\
        {\tt [7]} This line is not indented. What follows is
                another  line which is indented
                by 1.0~in.\\[1.0in]
        {\tt [8]} This is some more text.

        {\tt [9]} Let us show one more example: After the next
        line break, the line is indented by 1.0~in, but the text
        wraps around because it is longer than the remainder
        of the line:\\[1.0in]
```

```
                {\tt [10]} Some text, some more text, and even more.
        Then we have more text, and finally we are out of all
        the text we ever wanted to write. So that's it. Now we
        have some text that starts at the beginning of a line
        with square brackets like \\{}[1.0in] and it also works!
    }
$$
```

This input generated the following output:

[1] This is a sample of our new \\ macro. Let us see whether this all works now. The macro is not very difficult, so not much text is needed. Here we force a line break.
[2] And here is another one.
[3] And now let us force a line break where the following line is a little bit indented as it is here.
[4] So this is now indented.
[5] This is now indented by twice the amount of the paragraph indentation.
[6] This line is again indented by **\parindent**.
[7] This line is not indented. What follows is another line which is indented by 1.0 in.
[8] This is some more text.

[9] Let us show one more example: After the next line break, the line is indented by 1.0 in, but the text wraps around because it is longer than the remainder of the line:
[10] Some text, some more text, and even more. Then we have more text, and finally we are out of all the text we ever wanted to write. So that's it. Now we have some text that starts at the beginning of a line with square brackets like
[1.0in] and it also works!

10.11 \parfillskip

\parfillskip is a glue parameter. Its value determines the amount of horizontal glue which is inserted at the end of the last line of a paragraph; one could think of it as a special **\rightskip** glue for the last line of a paragraph (the regular **\rightskip** glue is still in effect, and in particular, it is also inserted at the end of the last line).

The default value of **\parfillskip** is `0pt plus 1fil`. This means that the glue can stretch as far as necessary to allow the last line to fall short of a full line; exactly what is normally intended to happen.

\parfillskip, if set to zero, causes TEX to generate a paragraph with a right justified last line (if left and right justified is the current paragraph layout). Let me discuss the cases where **\parfillskip** is set to zero:

1. When centering consecutive lines of text, the last line of the text should be centered too, and therefore, the special treatment of the last line must be disabled; see 11.6.2, p. 100.
2. When printing right justified lines, where the user programs the line breaks, the last line should also be right justified. Again **\parfillskip** must be disabled in this case. See 10.10.4, p. 41.

Let me now present a simple example. The same paragraph is printed twice, first with the default of **\parfillskip** in effect and then with this glue set to zero.

So, this is one more of these silly texts, but then on the other hand we do need examples, right? Sure, we do!

The same text with **\parfillskip = 0pt** added generates the following output:

So, this is one more of these silly texts, but then on the other hand we do need examples, right? Sure, we do!

10.12 Macro for Writing Paragraph Layout Parameters to the Logfile

The following macro source code file writes those parameters of TEX to the log file which have to do with line break and page break related computations.

$\mathcal{P}'$ • **wl-parcp.tip** •

```
\def\WritingParShape{%
    \wlog{\string\WritingParShape: begin}%
    \wlog{}%
    \wlog{\string\pretolerance: \the\pretolerance}%
    \wlog{\string\tolerance: \the\tolerance}%
    \wlog{\string\prevgraf: \the\prevgraf}%
    \wlog{}%
    \wlog{\string\parskip: \the\parskip}%
    \wlog{\string\baselineskip: \the\baselineskip}%
    \wlog{\string\lineskip: \the\lineskip}%
    \wlog{\string\lineskiplimit: \the\lineskiplimit}%
    \wlog{}%
    \wlog{\string\parindent: \the\parindent}%
    \wlog{\string\hsize: \the\hsize}%
```

```
    \wlog{\string\leftskip: \the\leftskip}%
    \wlog{\string\rightskip: \the\rightskip}%
    \wlog{\string\parfillskip: \the\parfillskip}%
    \wlog{\string\spaceskip: \the\spaceskip}%
    \wlog{\string\xspaceskip: \the\xspaceskip}%
    \wlog{}%
    \wlog{\string\hyphenpenalty: \the\hyphenpenalty}%
    \wlog{\string\exhyphenpenalty: \the\exhyphenpenalty}%
    \wlog{\string\lefthyphenmin: \the\lefthyphenmin}%
    \wlog{\string\righthyphenmin: \the\righthyphenmin}%
    \wlog{\string\lefthyphenmin: \the\lefthyphenmin}%
    \wlog{}%
    \wlog{\string\adjdemerits: \the\adjdemerits}%
    \wlog{\string\doublehyphendemerits: \the\doublehyphendemerits}%
    \wlog{\string\finalhyphendemerits: \the\finalhyphendemerits}%
    \wlog{\string\linepenalty: \the\linepenalty}%
    \wlog{\string\sfcode\string\`.: \the\sfcode`\.}%
    \wlog{\string\hyphenchar: \the\hyphenchar\tenrm}%
    \wlog{}%
    \wlog{\string\binoppenalty: \the\binoppenalty}%
    \wlog{\string\relpenalty: \the\relpenalty}%
    \wlog{}%
    \wlog{\string\everypar: \the\everypar}%
    \wlog{\string\WritingParShape: end}%
}
```

• End of `wl-parcp.tip` •

10.13 Summary

In this chapter we learned:

- Spaces and returns are equivalent when it comes to the typesetting of paragraphs. Multiple spaces in the input are reduced to simple spaces.
- TEX's default is to produce left and right justified lines. The line length of a paragraph is `\hsize` – `\leftskip` – `\rightskip`. The two glues `\leftskip` and `\rightskip` are, by default, zero.
- Paragraphs can be enclosed inside vboxes (which prevent page breaks in the middle of paragraphs) and those vboxes can be centered using display math mode.
- Normally a paragraph is started upon entering the first character of the text. Other ways to start a paragraph include mathematical equations, `\indent`, `\noindent`, `\hskip` and `\vrule`. The `\leavevmode` instruction is an explicit instruction to start a paragraph, which is most frequently used when paragraphs begin with an hbox.

- At the beginning of a paragraph, TEX generates the **\parskip** glue (an implicit vertical glue which is inserted before each paragraph), switches to horizontal mode, adds the paragraph indentation (amount determined by the dimension parameter **\parindent**), and evaluates **\everypar**.
- A paragraph is ended using **\par**. An easier way to discover the end of a paragraph on the terminal screen is by the empty line that follows some text. TEX translates an empty line to **\par**. **\par** is illegal inside mathematical equations and as part of arguments to a macro, unless the macro was defined using **\long**. The current values of **\baselineskip**, **\leftskip**, **\rightskip**, and so forth, are applied to the whole paragraph.
- **\leftskip** and **\rightskip** can be used to generate indented paragraphs. For instance, the macro **\narrower** changes both of these parameters to achieve indented paragraphs.
- The paragraph indentation is determined by the dimension parameter **\parindent**. A paragraph's indentation can be suppressed using **\noindent**, which also causes TEX to start a paragraph. A paragraph can be indented by a negative value to make the first line stick out.
- **\everypar** is a token parameter. TEX evaluates the register at the beginning of a paragraph after the **\parskip** glue is generated, and after TEX enters horizontal mode and generates the paragraph indentation.
- User-controlled line breaks (where line breaks in the output are generated from line breaks in the input) can be generated in any of the following three ways: 1) by generating short single line paragraphs (the macro **\obeylines** comes in handy in this case), 2) by forcing explicit line breaks within a paragraph using **\hfil\break**, and 3) by using hboxes directly.
- **\parfillskip** glue is inserted at the end of the last line of every paragraph. Its default value is `0pt plus 1fil`, and allows the last line of a paragraph to be shorter than all other lines.

11
Paragraphs, Part II

This chapter continues the discussion of paragraph processing in TEX. The main emphasis will be on lists. We will also tell how to generate headings in documents, and finally we will discuss hanging indentation.

11.1 Generating Lists

I will now show how lists can be typeset in TEX. *Lists* are texts, where paragraphs are indented (usually primarily on the left side) and those paragraphs are labeled with *item labels*, typically printed to the left of such paragraphs.

Many times item labels of one particular list are of *different length*, so one must decide how these item labels should be aligned with respect to each other. I will discuss *left* and *right* flush item labels. Left flush item labels are used for item labels consisting of text. Right flush item labels are used for item labels consisting of numbers, so that, for instance, labels "9." and "10." line up properly, by their last digit.

All together there are four different combinations of the placement of labels and their alignment, which are identified by the following two letter code:

1. LL. Item label on the left side of the paragraph, the item label itself is left flush.
2. LR. Item label on the left side of the paragraph, the item label itself is right flush.
3. RL. Item label on the right side of the paragraph, the item label itself is left flush.
4. RR. Item label on the right side of the paragraph, the item label itself is label right flush.

Fig. 11.1 on the next page shows the four possibilities. The labeling in this figure (as well as the definition of the names of some macros to be defined later) follows the code just established: the first letter identifies on which *side of the*

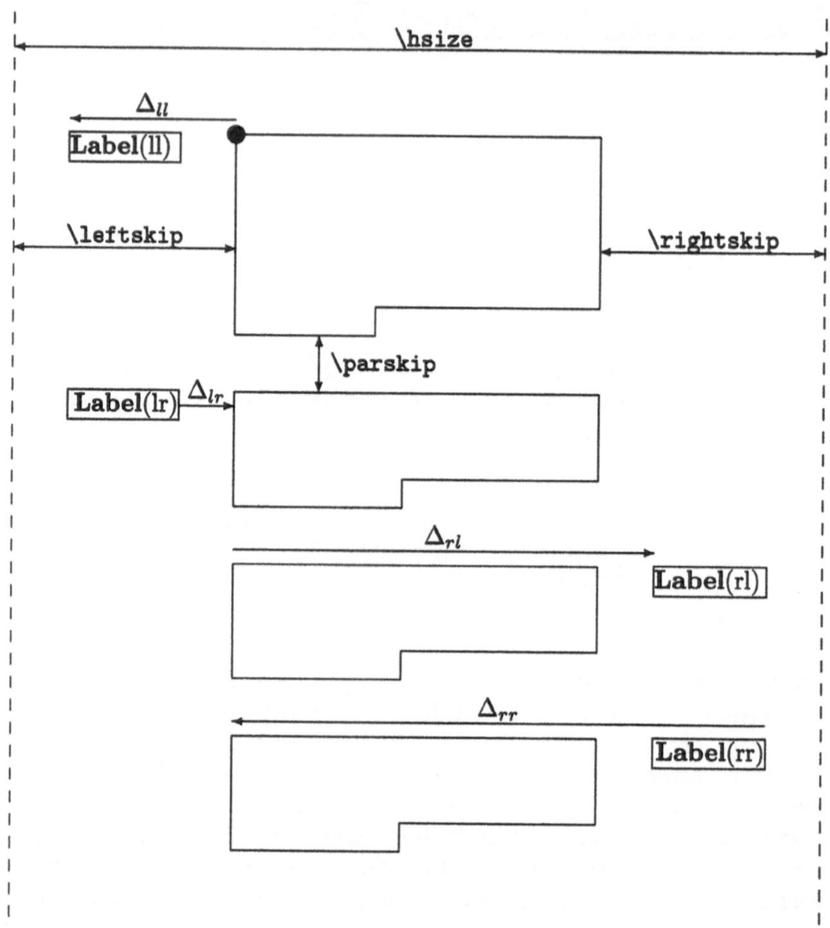

Figure 11.1. Paragraphs with labels.

paragraph the item label is printed, and the second code specifies *whether the item labels themselves are right flush or left flush* with respect to each other.

The various distances $\Delta_{x,y}$ are also indicated in Fig. 11.1 on this page. The user is responsible for choosing proper values for $\Delta_{x,y}$ as well as for `\leftskip` and `\rightskip`, which take the sizes of the labels printed later into account.

11.1.1 Item Labels Left of a Paragraph, Left Flush

Let me now show how you generate *left flush item labels* to the *left of a paragraph*. The following source code is used in the example:

```
\parskip = 12pt plus 2pt minus 1pt
\parindent = 20pt

{\tt [1]} This is some text preceding our label examples.
I will show here left flush labels which appear to the left
of a paragraph.

{%
    \leftskip = 1.5in
    \rightskip = 0.5in
    \noindent
    \hbox to 0pt{%
        \hskip -1.0in
        {\bf A label:}%
        \hfil
    }%
    {\tt [2]} This is fun what you can do with \TeX. In this case
    we have a short paragraph with a label and that's all we need to
    demonstrate what is going on.

    {\tt [3]} Here is a paragraph without a label. The paragraph is
    now indented. That's what we want.

    \noindent
    \hbox to 0pt{%
        \hskip -1.0in
        {\bf Another one:}%
        \hfil
    }%
    {\tt [4]} We have a paragraph with another label. Let it be
    with that. Note that we left an empty line before we started
    this paragraph.

    \noindent
    \hbox to 0pt{%
        \hskip -1.0in
        {\bf The last one:}%
        \hfil
    }%
    {\tt [5]} Now we have a paragraph with another label. We need
    such an example so you can see that indeed the labels are
    left flush.

}

{\tt [6]} And this is just some text at the end.
This text is outside the labels which we just demonstrated.
```

The output generated by the preceding source code reads as follows:

[1] This is some text preceding our label examples. I will show here left flush labels which appear to the left of a paragraph.

A label: [2] This is fun what you can do with TEX. In this case we have a short paragraph with a label and that's all we need to demonstrate what is going on.

[3] Here is a paragraph without a label. The paragraph is now indented. That's what we want.

Another one: [4] We have a paragraph with another label. Let it be with that. Note that we left an empty line before we started this paragraph.

The last one: [5] Now we have a paragraph with another label. We need such an example so you can see that indeed the labels are left flush.

[6] And this is just some text at the end. This text is outside the labels which we just demonstrated.

Understanding the following discussion is *essential*, because the technique used in the preceding example is used in all other list examples and macros. The changes necessary to accommodate different label types (label alignments) are minor.

1. `\leftskip` and `\rightskip` are used to indent the paragraph margins on both sides. Obviously, especially for labels to the left of a paragraph the value of `\leftskip` must be chosen with care, taking into account the width of the widest item label of a list and the desired distance of the item label from the text itself.
2. Before an item label is printed, a `\noindent` instruction is executed for *two* purposes:

   (a) The `\noindent`'s first function is to suppress the paragraph indentation. A paragraph with an item label should not be indented (disregarding the item label itself for now).

   (b) The `\noindent`'s second function is to *start a paragraph*. Explicitly start the paragraph to which this item label belongs is necessary, because the first item label of this paragraph is an hbox (the definition of this hbox starts on the following source code line), and hboxes do not start paragraphs. See 10.4.1, item 4, p. 11, for a detailed discussion of this issue.

   Note that because a paragraph is started *without* indentation, the current reference point is moved to the beginning of the first line of the paragraph. This position is the starting point for the printing of the item label.
3. Each item label is enclosed inside an hbox of zero width. This means that the item label is printed, but in the end the reference point as it was just

established at the beginning of the first line of the paragraph is *not* moved. See 6.5, p. I-180, for a discussion of zero width hboxes. Therefore, after the item label has been printed, TEX's current position is still the beginning of the first line. It is now possible to print the text of the paragraph "as if nothing had happened."

Let us now analyze the content of the zero width hbox. Regardless of the alignment type of the item label *and* where the item label occurs (to the left or to the right of a paragraph), this zero width hbox always has the same structure. Let me discuss here the case of a *left flush* label to the *left* of a paragraph as it was established through the preceding example.

This zero width hbox contains the following three items:

(a) *Horizontal glue to position to the beginning of where the item label is printed.* In the example horizontal glue in the amount of -1.0 in (`\hskip -1.0in`) moves the reference point 1.0 in to the left. Because `\leftskip` was set to 1.5 in before, the current horizontal position is now 0.5 in to the right from the ordinary text margin.

(b) *The text of the label.* The second item is the text of the label, which is printed now (in the example the label is printed in boldface).

(c) *Glue to position back to the original reference point.* `\hfil` glue that is glue which can stretch by an infinite amount is inserted now. This glue causes the reference point to move back to the original reference point (a zero width hbox is being constructed).

11.1.2 Item Labels Left of a Paragraph, Right Flush

Compared with the previous examples of left flush item labels very little changes with labels to the left of the list, but with the label item right flush. The only real difference is in the setup of the zero width hbox. The box has again three items in it which are slightly different now:

1. *Glue to position to the beginning of the label.* How far TEX has to go to the left before printing the label depends on the *length* of each label. We will therefore use glue which can *shrink* as far as necessary, as in `\hskip 0pt minus 1fil`. It is shorter to write `\hss`, although it is less precise, because `\hss` glue can also stretch (this would never be the case in the current setup).

   This step corresponds to item 3.a on this page and is slightly different in that the specified glue is an infinite glue rather than a glue of some specific distance.
2. *Print item label.* The item label is now printed. This step is identical to item 3.b on this page.
3. *Glue to position back to the original reference point.* Regardless of how long the item label was, the current reference point must be now moved to the right, by the same distance for all item labels. In the example we choose a distance of 30 pt. Going to the right by always the same distance brings TEX

back to where it started from. Because this dimension is the same regardless of the length of the item labels, all labels appear right flush.

This step corresponds to item 3.c on the previous page. It is different in that the distance moved to the right is always the same, and does not depend on the length of the label.

Here is an example of a list with labels that are right flush. First the source code:

```
\parskip = 12pt
\parindent = 20pt

{\tt [1]} This is some text preceding our label examples.
We will show here right flush labels which appear to the left
of a paragraph.

{%
    \leftskip = 1.5in
    \rightskip = 0.5in
    \noindent
    \hbox to 0pt{%
        \hss
        {\it 9:}%
        \hskip 30pt
    }%
    {\tt [2]} This is fun what you can do with \TeX. In our case
    we have a short paragraph with a label and that's all we need to
    demonstrate what is going on.

    {\tt [3]} Here is a paragraph without a label. That paragraph is
    now indented. But that's what we want.

    \noindent
    \hbox to 0pt{%
        \hss
        {\it 10:}%
        \hskip 30pt
    }%
    {\tt [4]} We now have a paragraph with another label.
    We need such an example so you can see that indeed the
    labels are right flush.

    \noindent
    \hbox to 0pt{%
        \hss
        {\it 100:}%
        \hskip 30pt
    }%
    {\tt [5]} Well here is just another example.

}

```

```
{\tt [6]} And this now is just some text at the end.
This text is outside the ``paragraph with labels area.''
```

The output generated by the preceding source code reads as follows:

`[1]` This is some text preceding our label examples. We will show here right flush labels which appear to the left of a paragraph.

9: `[2]` This is fun what you can do with TEX. In our case we have a short paragraph with a label and that's all we need to demonstrate what is going on.

`[3]` Here is a paragraph without a label. That paragraph is now indented. But that's what we want.

10: `[4]` We now have a paragraph with another label. We need such an example so you can see that indeed the labels are right flush.

100: `[5]` Well here is just another example.

`[6]` And this now is just some text at the end. This text is outside the "paragraph with labels area."

11.1.3 Macros for Printing Lists

Several macros based on the discussion of the previous subsections will aid in the typesetting of lists. Here is a brief overview of the macros I present.

1. Macros `\BeginAListX` and `\EndAListX` enclose a list. `\BeginAListX` sets up `\leftskip` and `\rightskip`; when `\EndAListX` is called, those changes are undone.

   The macros have a trailing `X` in their name to distinguish them from later following macros which are more powerful and general than the macros being offered here.

   The `\BeginAList` macro has the following three parameters:

   - `#1`. The relative increase of `\leftskip` (compared to its current value).
   - `#2`. The relative increase of `\rightskip` (compared to its current value).
   - `#3`. The value of $\Delta_{x,y}$. This dimension is always positive. You have to refer to Fig. 11.1, p. 50, for an interpretation of this value, because the interpretation of this value depends on the positioning and the alignment of the item labels.

Note that `#1` and `#2` are changes relative to `\leftskip` and `\rightskip`, because then it is easier to generate nested lists.

2. In addition to that the following macros will be defined (each of these macros has one parameter, `#1`, the text of the label to be printed). The codes used in the names of those macros are the codes defined before, that is, for instance, `\ItemLR` generates a left flush label to the right of a paragraph. See 11.1.3.2 on the next page for the definitions of the following four macros:

   (a) `\ItemLL`
   (b) `\ItemLR`
   (c) `\ItemRL`
   (d) `\ItemRR`

11.1.3.1 Macros To Set-Up the Indentation, `\BeginAListX`, `\EndAListX`

Here is the definition of macros `\BeginAListX` and `\EndAListX`.

$\mathcal{P}$ • par-lab.tip •

```
\catcode`\@ = 11
```

Store Δ_{xy} in the following dimension register.

```
\newdimen\Delta@XY
```

Keep track of the depth of nesting of `\BeginAListX` calls. Initial value is 0, each call of `\BeginAListX` increments this counter register by 1, each `\EndAListX` call decrements it by one.

```
\newcount\@LabeledParNesting
\@LabeledParNesting = 0
```

Here begins the definition of `\BeginAListX`. This macro has three parameters, which were explained before.

```
\def\BeginAListX #1#2#3{%
```

Finish the preceding paragraph, if it was not done already. Start a group to keep all changes local (the group will be terminated by `\EndAListX`).

```
    \par
    \bgroup
```

Now update `\leftskip` and `\rightskip`.

```
    \advance\leftskip by #1
    \advance\rightskip by #2
```

Increase the nesting counter.

```
    \advance\@LabeledParNesting by 1
```

Print message.

```
    \message{\string\BeginAListX: nesting level:
        \the\@LabeledParNesting}%
```

Set the label counter for the current list to zero.

```
        \@LabelCounter = 0
```

Save Δ_{xy}. Generate an error if it is negative.

```
        \Delta@XY = #3
        \ifdim\Delta@XY < 0pt
            \message{\string\BeginAListX: negative Delta{xy},
                made positive.}%
        \fi
    }
```

Next define macro `\EndAListX`, the macro which ends a group of labeled paragraphs. This macro has no parameters.

```
    \def\EndAListX{%
```

Finish the preceding paragraph (`\par` has no effect, if the call to `\EndAListX` was preceded by an empty line anyway). Undo all changes to `\leftskip`, `\rightskip` and so on by terminating the group started by `\BeginAListX`.

```
        \par
        \egroup
    }
```

11.1.3.2 Macros To Print the Labels

Macros `\ItemLL`, `\ItemLR`, `\ItemRL` and `\ItemRR` will print the labels. These macros are almost identical, with the exception of the zero width hbox which is constructed. Only the first macro `\ItemLL` is explained in detail; the explanation of the other three list-related macros focuses on the contents of the hboxes. All macros have one parameter, `#1`, the text of the label to be used.

`\ItemLL` must be applied between a pair of `\BeginLabelPar` and `\EndAListX` calls. A call to this macro generates a left justified label to the left of a paragraph.

```
    \def\ItemLL #1{%
```

Make sure the preceding paragraph is terminated.

```
        \par
```

Suppress the following paragraph's indentation. Start paragraph with a zero width hbox. Note the use of the "%" to suppress the end-of-line character after the opening curly brace of `\hbox to 0pt`. Then position to the beginning of the label, print the label and go back to the original reference point.

```
        \noindent
        \hbox to 0pt{%
            \hskip -\Delta@XY
            #1%
            \hfil
        }%
```

Any spaces immediately following a call to this macro should be suppressed; note that TEX is already in horizontal mode and therefore spaces are *not* ignored, but appear in the output. By inserting `\ignorespaces` (see 21.11, p. III-203) you can call this macro as follows:

```
\ItemLR{A Label}
    This is some text of a paragraph.
```

and the space token between the closing curly brace of `\ItemLR{A Label}` and the following text is ignored and does not translate into a space. If you omitted the `\ignorespaces` instruction, the first line of the list (ignoring the item label) would appear to be indented by a space.

Back to the code of `\ItemLL`; here is the promised `\ignorespaces`.

```
    \ignorespaces
}
```

Define macro `\ItemLR` to print a label to the left of a paragraph, with the label being right flush. The major change is to the zero width hbox which is now being built.

```
\def\ItemLR #1{%
    \par
    \noindent
    \hbox to 0pt{%
        \hss
        #1%
        \hskip\Delta@XY
    }%
    \ignorespaces
}
```

Define macro `\ItemRL` to print a label to the right of a paragraph, the label itself being left flush.

```
\def\ItemRL #1{%
    \par
    \noindent
    \hbox to 0pt{%
        \hskip\Delta@XY
        #1%
        \hss
    }%
    \ignorespaces
}
```

Define macro `\ItemRR` to print a label to the right of a paragraph, right justified.

```
\def\ItemRR #1{%
    \par
    \noindent
    \hbox to 0pt{%
        \hfil
        #1%
```

```
        \hskip -\Delta@XY
    }%
    \ignorespaces
}
\catcode'\@ = 12
```

• End of par-lab.tip •

11.1.4 An Example Using the Preceding Macros

Let us now look at an example using the preceding macros. The following source code is used in this example:

```
\input inputd.tip
\InputD{par-lab.tip}                          % 11.1.3.1, p. 56.
{
    \narrower
    This is some text preceding the example. Next the labeled
    paragraphs are started.
    \BeginAListX{1.0in}{0.5in}{10pt}
        \ItemLR{1.} This is great fun to me.
        \ItemLR{2.} This should also be fun to you.
        \ItemLR{3.} And so on \dots
        \ItemLR{4.} And here is more and more \dots
        \ItemLR{5.} And this is even more than more of more \dots
        \ItemLR{$\ldots$} And this is even more than more of more.
        \ItemLR{8.} And so on \dots
        \ItemLR{9.} Another silly text.
        \ItemLR{10.} The last text to show the alignment
            of labels.
    \EndAListX
    This is some text following the labeled paragraphs.

}
```

This source code generates the following output:

This is some text preceding the example. Next the labeled paragraphs are started.

1. This is great fun to me.
2. This should also be fun to you.
3. And so on ...
4. And here is more and more ...
5. And this is even more than more of more ...

... And this is even more than more of more.

8. And so on ...
9. Another silly text.
10. The last text to show the alignment of labels.

This is some text following the labeled paragraphs.

11.1.5 Selecting Proper Dimensions

It is obvious from the preceding macros that one must have an idea of what the width of the widest label in a group of labels is going to be to set up all dimensions properly when **\BeginAListX** is called. For labels consisting of numbers this is rather straightforward because such numbers will rarely be longer than two digits, so a number with two digits can be taken as the longest possible number.

For item labels with text one has to know the width of the widest item label because this value goes into the proper computation of $\Delta_{x,y}$ and **\leftskip**.

The width of the widest label can be determined easily using a box register (see 4.5, p. I-95). Let me present an example where Δ_{xy} is determined as 10 pt plus the width of the widest label.

```
\dimen0 = 10pt
\setbox0 = \hbox{Widest item label}
\advance\dimen0 by \wd0
\BeginAListX{...}{...}{\dimen0}
```

11.1.6 How To Handle Item Labels That Are Too Wide

Now I will discuss what can be done with item labels that are too wide. Note that it is fairly easy to discover whether an item label is too wide or not. Proceed as shown in the preceding Subsection, measuring the length by storing the label in an hbox and assigning it to a box register.

You can handle oversized labels in the following ways:

1. Simply *ignore the problem* (the world will not disappear, because you do that). The current version of the preceding macros will cause the item label to run into the text.
2. Print a warning message, but do nothing special (certainly better than the preceding solution). This solution is discussed in 11.1.6.1 on this page.
3. Force a line break after the item label so that the paragraph of this item label starts on the next line. This solution is shown further below.
4. Start the first line of the paragraph's text further to the right than usual so there is sufficient room for the item label, also discussed below.
5. Move the label itself further to the left.

11.1.6.1 A New \ItemLL, Called \ItemLLOne

Here is a new definition of **\ItemLL** called **\ItemLLOne** that will print a message if a label is too wide. This macro, as before, generates a left justified label to the left of a paragraph. It does not try to correct this mistake in any way.

P • **itemltw.tip** •

```
\def\ItemLLOne #1{%
    \par
```

The following four lines were added compared to the original definition.

```
    \setbox 0 = \hbox{#1}%
    \ifdim\wd0 > \Delta@XY
        \message{\string\ItemLL: warning, label "#1" too wide.}%
    \fi
```

From here on it's like before.

```
    \noindent
    \hbox to 0pt{%
        \hskip -\Delta@XY
        #1%
        \hfil
    }%
    \ignorespaces
}
```

• End of `itemltw.tip` •

11.1.6.2 Forcing a Line Break If Label Too Wide, `\ItemLLTwo`

Here is yet another version of `\ItemLL`, called `\ItemLLTwo` that, if the label is too wide, generates a line break right after the item label is printed.

• **label-ll-w1.tip** •

```
\catcode`\@ = 11
\def\ItemLLTwo #1{%
    \par
    \noindent
    \hbox to 0pt{%
        \hskip -\Delta@XY
        #1%
```

Use `\hss` glue. This glue actually shrinks backwards, if the item label is too wide. No message about an overfull hbox is generated.

```
        \hss
    }%
```

In case the item label is too wide, generate a line break now which causes the paragraph's text to start on the next line. Also print a warning message. The generation of line breaks in a paragraph using `\hfil\break` is discussed in 10.10.1, p. 36.

```
    \setbox 0 = \hbox{#1}%
    \ifdim\wd0 > \Delta@XY
        \hfil\break
        \message{\string\ItemLL: Forced line break
```

```
                inserted, item label "#1" too wide.}%
        \fi
```

Continue as before.

```
        \ignorespaces
}
\catcode'\@ = 12
```

• End of `label-11-w1.tip` •

11.1.7 The Automatic Numbering of Lists

Based on the preceding macro definitions let me now present a macro to automatically number a list. The `\ItemLR` macro is used by the macro `\CountingItem` which is defined here. `\CountingItem` has no parameter, because the numbering of the list is done by this macro and therefore there is no need to provide a label.

$\mathcal{P}$ • `par-lcl.tip` •

```
\InputD{par-lab.tip}                        % 11.1.3.1, p. 56.
\catcode'\@ = 11
```

Declare a counter for item labels with numbers.

```
\newcount\@LabelCounter
\def\CountingItem{%
```

Increment the label counter first. The label counter was defined previously.

```
        \advance\@LabelCounter by 1
```

This macro supports three levels of nesting: the outermost item labels use arabic numerals, the next level uses lowercase letters enclosed in parenthesis, and the third level uses lowercase roman numerals. The item label is now generated.

```
        \ifcase\@LabeledParNesting
```

There is no level 0 numbering.

```
        \or
            \ItemLR{\the\@LabelCounter.}%
        \or
```

For labels using letters the label counter value must be increased by the character code of `a` (which is done by the `'a` construct). Instead of this approach the macro `\alph` of 3.4.1, p. I-68, could have been used.

```
            \ItemLR{%
                \advance\@LabelCounter by '\a
                (\char\@LabelCounter)%
            }%
        \or
            \ItemLR{\romannumeral\@LabelCounter.}%
```

Generate an error if lists are too deeply nested.

```
    \else
        \errmessage{\string\CountingItem: nesting level up to 3
            supported only.}%
    \fi
}
\catcode'\@ = 12
```

• End of par-lcl.tip •

Observe the following example. This input was used for this example:

```
\InputD{par-lcl.tip}                        % 11.1.7, p. 62.

This is some text preceding the other text which is some wonderful
other and so on text.
This is some text preceding the other text which is some wonderful
other and so on text.
\BeginAListX{30pt}{0.0in}{10pt}
    \CountingItem This is some fun text.
    \CountingItem This is also some fun text.
        \BeginAListX{30pt}{0.0in}{10pt}
            \CountingItem This is some fun text.
            \CountingItem This is also some fun text.
            \CountingItem This is some fun text.
            \CountingItem This is also some fun text.
            \CountingItem This is some fun text.
            \CountingItem This is also some fun text.
        \EndAListX
    \CountingItem This is also some fun text.
    \CountingItem This is also some fun text.
        \BeginAListX{30pt}{0.0in}{10pt}
            \CountingItem This is some fun text.
            \CountingItem This is also some fun text.
            \CountingItem This is some fun text.
                \BeginAListX{30pt}{0.0in}{10pt}
                    \CountingItem This is some fun text.
                    \CountingItem This is also some fun text.
                    \CountingItem This is some fun text.
                    \CountingItem This is also some fun text.
                    \CountingItem This is some fun text.
                \EndAListX
            \CountingItem This is also some fun text.
            \CountingItem This is some fun text.
            \CountingItem This is also some fun text.
        \EndAListX
    \CountingItem This is also some fun text.
    \CountingItem This is also some fun text.
\EndAListX
```

The output generated by the preceding source code can be found in Figure 11.2 on the next page.

This is some text preceding the other text which is some wonderful other and so on text. This is some text preceding the other text which is some wonderful other and so on text.

1. This is some fun text.
2. This is also some fun text.
    - (b) This is some fun text.
    - (c) This is also some fun text.
    - (d) This is some fun text.
    - (e) This is also some fun text.
    - (f) This is some fun text.
    - (g) This is also some fun text.
3. This is also some fun text.
4. This is also some fun text.
    - (b) This is some fun text.
    - (c) This is also some fun text.
    - (d) This is some fun text.
        - i. This is some fun text.
        - ii. This is also some fun text.
        - iii. This is some fun text.
        - iv. This is also some fun text.
        - v. This is some fun text.
    - (e) This is also some fun text.
    - (f) This is some fun text.
    - (g) This is also some fun text.
5. This is also some fun text.
6. This is also some fun text.

Figure 11.2. Numbered list example output.

11.1.8 An Interesting Variation of the List Problem

A variation of the list problem is to generate a page layout as it can be found in some manuals and catalogues: section numbers form the first column, titles of those sections the second column, and the third column contains the text itself. This problem can be solved easily if one looks at the section numbers and titles as item labels of a list, where the list's text is the main text of the document.

For example:

• **ex-labpar.tip** •

Change `\leftskip`. `\rightskip` remains zero.

```
\leftskip = 2in
\rightskip = 0pt
```

Set-up `\parskip`.

```
\parskip = 12pt plus 2pt minus 1pt
```

Next define macro `\Sec`, which has two parameters:

- #1. The section number.
- #2. The section title.

The beginning of such a macro should be familiar to you by now: terminate the preceding paragraph, if necessary, and start a non-indented paragraph with a zero width hbox.

```
\def\Sec #1#2{%
    \par
    \noindent
    \hbox to 0pt{%
```

Print number and title of section in boldface.

```
        \bf
```

Go back to the left margin, print the number there, followed by the title 0.5 in to the left of the margin.

```
        \hskip -\leftskip
        \hbox to 0.5in{%
            #1%
            \hfil
        }%
```

Print the title.

```
        #2%
        \hfil
    }%
    \ignorespaces
}
```

Now show some examples. The text is entered in a way similar to the way it will be printed, although this is not really necessary as far as the final output is concerned.

```
\Sec{1}{Introduction}   So here is a section
                        where we show all the ideas
                        about having double labels
                        to the left of a paragraph.
\Sec{1.1}{Acknowledgments} And you really want to be careful
                        so your entries are not too wide.
                        You may get an overfull box in those cases.

                        Now this paragraph is indented. Remember
                        that the macro {\tt\string\Sec} used in the
                        previous paragraphs has a
                        {\tt\string\noindent} in the beginning.
\Sec{1.2}{More Stuff}   And here we can have some more stuff.
                        Really more than ever before.
                        And here we can have some more stuff.
                        Really more than ever before.
\Sec{1.3}{That's It}    Yes, that's it.
```

• End of `ex-labpar.tip` •

1	**Introduction**	So here is a section where we show all the ideas about having double labels to the left of a paragraph.
1.1	**Acknowledgments**	And you really want to be careful so your entries are not too wide. You may get an overfull box in those cases.
		Now this paragraph is indented. Remember that the macro `\Sec` used in the previous paragraphs has a `\noindent` in the beginning.
1.2	**More Stuff**	And here we can have some more stuff. Really more than ever before. And here we can have some more stuff. Really more than ever before.
1.3	**That's It**	Yes, that's it.

Figure 11.3. Sample output of list printing variation problem.

The output generated by the preceding source code can be found in Figure 11.3 on this page.

11.2 Improved Macros for Lists

Let us now examine some improved macros for the printing of lists. The main problem with the macros offered so far is that there is no control over the vertical spacing between list elements or before and after lists. On the other hand this is necessary, because you want to avoid making the user manually insert `\vskips` or related macros in a text to fix up the vertical spacing.

To understand these macros we must differentiate between two types of paragraphs: labeled paragraphs (with item labels) and unlabeled paragraphs (without item labels). For instance, if you look at item 1. of the following list, then the labeled paragraph is followed by an unlabeled one ("For symmetry...").

The following three glue values must be specified by the user. These values describe completely the vertical distance between any pair of consecutive paragraphs in a list. Note that glue values rather than vertical dimensions are used, because the user might provide some stretchability and shrinkability with those distances for TeX to be able to adjust the page lengths.

1. The vertical distance between the text preceding the very first item labeled and the beginning of this paragraph.

   For symmetry the same vertical space is inserted between the last paragraph of a list and the following text.

2. The distance between an unlabeled paragraph and a labeled paragraph.
3. The distance between a labeled paragraph and an unlabeled paragraph. The same value is applied between two unlabeled paragraphs.

11.2.1 The Macro Source Code

Here is the macro source code of the macros for lists.

$\mathcal{P}'$ • parv-1.tip •

```
\catcode`\@ = 11
```

Store Δ_{xy} in the following dimension register.

```
\newdimen \Delta@XY
```

Store the vertical space to be inserted before the whole list and after the whole list in the following register.

```
\newskip\@ParListBeforeAfter
```

Store the amount of vertical space between `\Item..` calls here.

```
\newskip\@ParListBetweenLabels
```

Store the amount of vertical space after a labeled and before an unlabeled paragraph here.

```
\newskip\@ParListAfterLabel
```

Keep track of the depth of the nestings of `\BeginAList` calls.

```
\newcount\@LabeledParNesting    \@LabeledParNesting = 0
```

Declare a counter for labels consisting of numbers like (1), (2), ..., or for that matter also of letters (a), (b), (c) etc.

```
\newcount\@LabelCounter
```

Here is the definition of `\BeginAList`. This macro has the following seven parameters:

- `#1`. The relative change of `\leftskip`. This amount therefore controls the horizontal position of the left margin of the text in a list (excluding item labels).
- `#2`. The relative change to `\rightskip`. This amount therefore controls the horizontal position of the right margin of the text of a list.
- `#3`. Δ_{xy}, always positive.
- `#4`. Vertical glue to be added before the very first and after the very last paragraph of a list.
- `#5`. Vertical glue to be added before each labeled paragraph with the exception of the very first labeled paragraph. The glue preceding the very first labeled paragraph is determined by `#4`.

- #6. Vertical glue to be used between a labeled and an unlabeled paragraph or between two unlabeled paragraphs.
- #7. The paragraph indentation used for unlabeled paragraphs.

```
\def\BeginAList #1#2#3#4#5#6#7{%
```

Finish the preceding paragraph, then start a group to keep all changes local (the group started here is later terminated by `\EndAList`).

```
    \par
    \bgroup
```

Now update `\leftskip` and `\rightskip`. Increase the nesting counter.

```
    \advance\leftskip by #1
    \advance\rightskip by #2
    \advance \@LabeledParNesting by 1
```

Set the counter of labels within the current list to zero. Because grouping is in effect there are no conflicts if the current list is contained in another list. Save Δ_{xy} and generate an error if it is negative.

```
    \@LabelCounter = 0
    \Delta@XY = #3
    \ifdim\Delta@XY < 0pt
        \errmessage{\string\BeginAList: negative Delta{xy},
            made positive.}%
    \fi
```

Save all the other parameters.

```
    \@ParListBeforeAfter = #4
    \@ParListBetweenLabels = #5
    \@ParListAfterLabel = #6
    \SetParIndent{#7}
}
```

Now define macro `\EndAList`, a macro which ends a group of labeled paragraphs. This macro has no parameters. Finish current paragraph (if not yet finished) and undo all changes to `\leftskip`, `\rightskip`, `\parindent`, by ending the group previously started.

```
\def\EndAList{%
    \par
    \vskip\@ParListBeforeAfter
    \egroup
}
```

11.2.2 The Definition of the Label-Generating Macros

Macros to generate item labels can be helpful, and then go with the preceding macros `\BeginAList` and `\EndAList`. First define macro `\@GenLabel` (generic label) which is used by all the subsequent macros. This macro's only parameter, `#1`, is the content of the zero width hbox which is used to print the label itself.

```
\def\@GenLabel #1{%
```

Make sure the previous paragraph is finished.

```
      \par
```

Advance the label counter **\@LabelCounter** by 1 to determine whether this is the very first labeled paragraph or not, because this has an influence on the vertical spacing. All vertical spacing is done by setting **\parskip** properly.

```
      \advance\@LabelCounter by 1
      \ifnum\@LabelCounter = 1
          \parskip = \@ParListBeforeAfter
      \else
          \parskip = \@ParListBetweenLabels
      \fi
```

Suppress the paragraph indentation and start the paragraph with a zero width hbox. Note the use of the "`%`" to suppress the end-of-line character after the opening curly brace of **\hbox to 0pt**. Then position to the beginning of the label, print the label and go back to the original reference point.

```
      \noindent
      \hbox to 0pt{#1}%
```

Now set **\parskip** to the value that is applied between a labeled and a following unlabeled paragraph. This value of **\parskip** may never be used, if the current paragraph is immediately followed by another labeled paragraph. Note that the assigned **\parskip** glue is also used to determine the distance between two consecutive unlabeled paragraphs. The value of **\parskip** as set here does *not* apply to the current paragraph, because TeX has already entered horizontal mode when the preceding **\noindent** was executed. See 12.2, p. 110, for details on this timing issue.

```
      \parskip = \@ParListAfterLabel
```

Any spaces immediately following a call to this macro should be suppressed (TeX is already in horizontal mode).

```
      \ignorespaces
}
```

Now define macro **\ItemLL**. This macro should be called after a **\BeginAList** to generate a left justified item label to the left of a paragraph. This macro has one parameter, **#1**, the text of the item label.

```
\def\ItemLL #1{%
    \@GenLabel{%
        \hskip -\Delta@XY
        #1%
        \hfil
    }%
}
```

Define macro `\ItemLR` to print an item label to the left of a paragraph, with the item label being right flush. Also this macro has one parameter, `#1`, the text of the item label.

```
\def\ItemLR #1{%
    \@GenLabel{%
        \hss
        #1%
        \hskip\Delta@XY
    }%
}
```

Define macro `\ItemRL` to print an item label to the right of a paragraph, with the item label being left flush. The text of the label is this macro's only parameter.

```
\def\ItemRL #1{%
    \@GenLabel{%
        \hskip\Delta@XY
        #1%
        \hss
    }%
}
```

Now define `\ItemRR` to generate an item label to the right of a paragraph with the item label itself being right flush. The text of this label is this macros only parameter.

```
\def\ItemRR #1{%
    \@GenLabel{%
        \hfil
        #1%
        \hskip -\Delta@XY
    }%
}
\catcode`\@ = 12
```

• End of `parv-1.tip` •

11.2.3 An Example Using the Preceding Macros

Let me present an example application of the preceding macros. The following input is used in it.

```
\input inputd.tip
\InputD{parv-1.tip}                          % 11.2.1, p. 67.

\leftskip = 0.5in
\rightskip = 0.5in

    This is some text preceding the labeled paragraphs. This text
is written here so you can see the vertical glue added by
```

```
the above macros between the text preceding the labeled
paragraphs and the first labeled paragraph.
\BeginAList{0.5in}{0pt}{0.25in}{12pt plus 2pt minus 2pt}%
    {6pt plus 1pt minus 1pt}{3pt plus 1pt minus 1pt}{20pt}
    \ItemLR{\tt \#1}
        This is the very first parameter. Let me tell you that.
        This parameter must be explained. Yes, so let me explain it.

            Interestingly enough a second paragraph is needed for the
        explanation, but you probably will not care about any detail
        described here.
    \ItemLR{\tt \#2}
        This is the very second parameter. Let me tell you that. Now
        the second parameter needs a little more detailed
        explanation, as far as I can tell you. The explanation, in
        fact, is so detailed that a new paragraph must be started.

            Here is this new paragraph.

            Here is the second new paragraph.  Much more is needed
        here, so let me go on with it. Nothing is wrong, but there
        is more explanation needed. So let me finish this now.
    \ItemLR{\tt \#3}
        This is the third parameter. Let me tell you that.
    \ItemLR{\tt \#4}
        This is the fourth parameter. Let me tell you that.
    \ItemLR{\tt \#5}
        This is the fifth and last parameter.
\EndAList
    This is some text following the last labeled paragraphs.
This text is written here so you can see the vertical space
added by the above macros between the text preceding the labeled
paragraphs and the first labeled paragraph.
```

This input generates the following output:

This is some text preceding the labeled paragraphs. This text is written here so you can see the vertical glue added by the above macros between the text preceding the labeled paragraphs and the first labeled paragraph.

#1 This is the very first parameter. Let me tell you that. This parameter must be explained. Yes, so let me explain it.

Interestingly enough a second paragraph is needed for the explanation, but you probably will not care about any detail described here.

#2 This is the very second parameter. Let me tell you that. Now the second parameter needs a little more detailed explanation, as far as I can tell you. The explanation, in fact, is so detailed that a new paragraph must be started.

Here is this new paragraph.

Here is the second new paragraph. Much more is needed here, so let me go on with it. Nothing is wrong, but there is more explanation needed. So let me finish this now.

#3 This is the third parameter. Let me tell you that.

#4 This is the fourth parameter. Let me tell you that.

#5 This is the fifth and last parameter.

This is some text following the last labeled paragraphs. This text is written here so you can see the vertical space added by the above macros between the text preceding the labeled paragraphs and the first labeled paragraph.

11.3 General List Macros

Based on the preceding discussion I will now offer macros which allow for automatically enumerated lists. These macros are set up to conform to the guidelines for this series, but a modification to fit your own needs should be straightforward. Note that I did *not* separate the layout information contained in the following macros and the macros themselves by introducing style files (see 27.6.5, p. III-437, for details on style files).

11.3.1 Numbered Lists

First let me deal with *numbered lists*, that is lists which are numbered, for instance, 1., 2., 3., and so forth. The macros below support nested lists, meaning numbered lists within other numbered lists.

$\mathcal{P}'$ • **enumlist.tip** •

```
\InputD{parv-1.tip}                          % 11.2.1, p. 67.
\InputD{counters.tip}                        % 3.4.2, p. I-71.
\catcode`\@ = 11
```

One needs to keep track of how deep such lists are nested, because depending on the nesting level different enumeration schemes are used.

```
\newcount\@EnumerateListDepth
\@EnumerateListDepth = 0
```

Declare four counters for the enumerated lists now. If you don't like the way lists are enumerated, use the **\ReassignCounter** macro of 3.4.3, p. I-72, to modify those definitions.

The outermost level uses arabic numerals, followed by a period.

```
\NewCounter{Enumerate1}{\arabic}%
    {\TheCounter{Enumerate1}.}{\TheCounter{Enumerate1}}
```

Next level: lower case letters, enclosed in parentheses.

```
\NewCounter{Enumerate2}{\alph}%
    {(\TheCounter{Enumerate2})}%
    {\TheCounter{Enumerate1}.\TheCounter{Enumerate2}}
```

Next level: roman numerals, lower case, followed by a period.

```
\NewCounter{Enumerate3}{\roman}%
    {\TheCounter{Enumerate3}.}%
    {\RefCounter{Enumerate2}.\TheCounter{Enumerate3}}
```

Last level: print uppercase letters, followed by a period.

```
\NewCounter{Enumerate4}{\Alph}%
    {\TheCounter{Enumerate4}.}%
    {\RefCounter{Enumerate3}.\TheCounter{Enumerate4}}
```

Now define macro **\BeginEnumerate** which starts an enumerated list. This macro is defined so that enumerated lists can occur inside other lists.

```
\def\BeginEnumerate{%
```

Start a group, and increment the nesting counter.

```
    \begingroup
    \global\advance\@EnumerateListDepth by 1
```

Now execute the code which corresponds to the current level of nesting.

```
    \ifcase\@EnumerateListDepth
```

There is no level 0.

```
        \errmessage{\string\BeginEnumerate: no level
            zero.}%
    \or
```

Level 1. Reset the counter for this level.

```
        \SetCounter{Enumerate1}{0}%
```

Call another macro which sets up all the dimensions, etc., for lists of that level.

```
        \@BeginEnumerateLevelOne
```

Define macro **\Label** so that a cross-reference to the current item label is generated, if **\Label** is used after an **\Item..** call. Cross-referencing is discussed in 19.3, p. III-85.

```
        \def\Label ##1{\@Label{##1}{\RefCounter{Enumerate1}}{1}}%
```

The definition of **\Item**, the macro to generate an item label, depends on the nesting level, because the proper counter to be used depends on the level of nesting. So you will see four different definitions of `\Item`, each for one of the four levels.

```
        \def\Item{%
            \StepCounter{Enumerate1}%
            \ItemLR{\PrintCounter{Enumerate1}}%
        }%
    \or
```

The code for the next level, level 2, follows next.

```
        \SetCounter{Enumerate2}{0}%
        \@BeginEnumerateLevelTwo
        \def\Label ##1{\@Label{##1}{\RefCounter{Enumerate2}}{1}}%
        \def\Item{%
            \StepCounter{Enumerate2}%
            \ItemLR{\PrintCounter{Enumerate2}}%
        }%
    \or
```

The code for level 3 follows next.

```
        \SetCounter{Enumerate3}{0}%
        \@BeginEnumerateLevelThree
        \def\Label ##1{\@Label{##1}{\RefCounter{Enumerate3}}{1}}%
        \def\Item{%
            \StepCounter{Enumerate3}%
            \ItemLR{\PrintCounter{Enumerate3}}%
        }%
    \or
```

The code for level 4 is last.

```
        \SetCounter{Enumerate4}{0}%
        \@BeginEnumerateLevelFour
        \def\Label ##1{\@Label{##1}{\RefCounter{Enumerate4}}{1}}%
        \def\Item{%
            \StepCounter{Enumerate4}%
            \ItemLR{\PrintCounter{Enumerate4}}%
        }%
```

In case the level is 5 or larger, an error is generated. Obviously it would be very simple to extend the preceding macro to allow for even more deeply nested lists.

```
    \else
        \errmessage{\string\BeginEnumerate: maximum
            nesting level of 4 exceeded.}%
    \fi
}
```

Next define macro `\EndEnumerate` which ends an enumerated list.

```
\def\EndEnumerate{%
    \EndAList
```

End the group started by `\BeginEnumerate`. Decrease the nesting counter.

```
    \endgroup
```

```
        \global\advance\@EnumerateListDepth by -1
}
\catcode`\@ = 12
```

• End of enumlist.tip •

11.3.2 Numbered Lists in This Series

Here is the macro source file `ts-enum.tip` that initializes enumerated list printing to conform to the style specifications of this publisher; see Guidelines (1990). The following macros are loaded to print all enumerated lists in this series.

$\mathcal{P}'$ • ts-enum.tip •

```
\InputD{enumlist.tip}                     % 11.3.1, p. 72.
\InputD{ts-fonts.tip}                     % 16.1.9, p. 271.
\InputD{widestc.tip}                      % 27.1.8.4, p. III-415.
\catcode`\@ = 11
```

Now compute the indentation of numbered lists from the left margin. To do so we need to add the width of the widest digit (using **\FindWidestChar**), the width of the period, and 1 em (1 em is the required space between the label and the labeled paragraph).

```
\newdimen\@WidthLevelOneLabels
\FindWidestChar{\@WidthLevelOneLabels}{\normalsize\rm}%
    {`\0}{`\9}
\setbox 0 = \hbox{.\hskip 1em}
\advance\@WidthLevelOneLabels by \wd0
```

The **\@BeginEnumeratedLevelX** macro all need to call **\BeginAList**. The main purpose of those macros is to set up dimensions properly.

```
\def\@BeginEnumerateLevelOne{%
    \BeginAList{\@WidthLevelOneLabels}{0pt}{1em}%
        {12pt}{0pt}%
        {0pt}{15pt}%
}
```

For enumerated lists of level 2 lowercase letters are used, enclosed in parentheses. Otherwise the same spacing rule as above is used. The search for the widest character is restricted to characters a–j, because lists are usually not numbered any further.

```
\newdimen\@WidthLevelTwoLabels
\FindWidestChar{\@WidthLevelTwoLabels}{\normalsize\rm}%
    {`\a}{`\j}
\setbox 0 = \hbox{()\hskip 1em}
\advance\@WidthLevelTwoLabels by \wd0
\def\@BeginEnumerateLevelTwo{%
    \BeginAList{\@WidthLevelTwoLabels}{0pt}{10pt}%
        {6pt}{0pt}%
```

```
            {0pt}{15pt}%
}
```

Do the same for lists of level 3. Let's assume here that "viii" is the widest roman numeral ever generated.

```
\newdimen\@WidthLevelThreeLabels
\setbox 0 = \hbox{viii.\hskip 1em}
\@WidthLevelThreeLabels = \wd0
\def\@BeginEnumerateLevelThree{%
      \BeginAList{\@WidthLevelThreeLabels}{0pt}{10pt}%
            {6pt}{0pt}%
            {0pt}{15pt}%
}
```

Repeat the same for lists of level 4. Width determination is similar to the level 2 width determination.

```
\newdimen\@WidthLevelFourLabels
\FindWidestChar{\@WidthLevelFourLabels}{\normalsize\rm}%
      {'\A}{'\J}
\setbox0 = \hbox{()\hskip 1em}
\advance\@WidthLevelFourLabels by \wd0
\def\@BeginEnumerateLevelFour{%
      \BeginAList{\@WidthLevelFourLabels}{0pt}{10pt}%
            {6pt}{0pt}%
            {0pt}{15pt}%
}
```

In this series I had a few lists (the list starting on p. I-16 is an example), where on the outermost level (level 1) the numbering goes beyond 9, so the longest item label does not consist of one, but of two digits. For that purpose I defined the following macro `\EnumerateLevelOneExtended`. When this macro is invoked, level 1 enumerated lists allocate more space for labels.

This macro is usually applied within a group *and* before `\BeginEnumerate` is called. After the associated group ends, level 1 enumerated lists continue as before.

```
\def\EnumerateLevelOneExtended{%
      \par
      \FindWidestChar{\@WidthLevelOneLabels}{\normalsize\rm}%
      {'\0}{'\9}
```

The following line has been added compared to the code which computes this dimension for regular level 1 lists. This stores the width of twice the widest digit into `\@WidthLevelOneLabels`. Note that in most fonts all digits have the same width and therefore there is no real need to compute the maximum width of all 10 digits. Therefore the preceding two source code lines and the following three source code lines could have been replaced by `\setbox 0 = \hbox{00.\hskip 1em}` followed by assigning `\wd0` to `\@WidthLevelOneLabels`.

```
      \multiply\@WidthLevelOneLabels by 2
```

As before take into account the space occupied by a period plus the white space between the period and the text of the paragraph.

```
        \setbox 0 = \hbox{.\hskip 1em}%
        \advance\@WidthLevelOneLabels by \wd0
}
\catcode`\@ = 12
```

• End of **ts-enum.tip** •

See 11.3.5, p. 79, for an example application of enumerated lists (or just look almost anywhere in this series, if you don't care about the source code).

11.3.3 Itemized Lists, \BeginItemize, \EndItemize

The next type of lists is called *itemized lists*, in which no numbers but bullets (•) and other characters are used. The macro source code below very much resembles the macro source code of the preceding two Subsections.

$\mathcal{P}'$ • **itemizel.tip** •

```
\InputD{parv-1.tip}                         % 11.2.1, p. 67.
\catcode`\@ = 11
```

Again one needs to keep track of how deep such lists are nested.

```
\newcount\@ItemListDepth
\@ItemListDepth = 0
```

Now define macro **\BeginItemize** which starts an itemized list.

```
\def\BeginItemize{%
    \begingroup
    \global\advance\@ItemListDepth by 1
```

Outermost lists use a bullet (•).

```
    \ifcase\@ItemListDepth\or
        \@BeginItemizeLevelOne
        \def\Item{\ItemLL{$\bullet$}}%
    \or
```

On the next leven en-dashes are used.

```
        \@BeginItemizeLevelTwo
        \def\Item{\ItemLL{--}}%
    \or
```

Level 3 uses asterisks.

```
        \@BeginItemizeLevelThree
        \def\Item{\ItemLL{*}}%
    \or
```

Finally, on level 4, there are plus signs.

```
        \@BeginItemizeLevelFour
        \def\Item{\ItemLL{+}}%
```

```
    \else
        \errmessage{\string\BeginItemize: maximum nesting of
            4 exceeded.}%
    \fi
}
```

The following definition of `\EndItemize` is almost identical to the definition of `\EndEnumerate`.

```
\def\EndItemize{%
    \EndAList
    \endgroup
    \global\advance\@ItemListDepth by -1
}
\catcode`\@ = 12
```

• End of `itemizel.tip` •

11.3.4 Itemized Lists for This Series

The following source code does the customization of the dimensions of itemized lists. The code is very similar to the preceding code dealing with the enumerated environment. The following code is used for itemized lists in this series.

$\mathcal{P}'$ • `ts-itize.tip` •

```
\InputD{itemizel.tip}                       % 11.3.3, p. 77.
\InputD{ts-fonts.tip}                       % 16.1.9, p. 271.
\catcode`\@ = 11
```

Level 1.

```
\def\@BeginItemizeLevelOne{%
    \BeginAList
        {15pt}{0pt}{15pt}%
        {12pt}{0pt}%
        {0pt}{15pt}%
}
```

Level 2.

```
\def\@BeginItemizeLevelTwo{%
    \BeginAList
        {15pt}{0pt}{15pt}%
        {6pt}{0pt}%
        {0pt}{15pt}%
}
```

Level 3.

```
\def\@BeginItemizeLevelThree{%
    \BeginAList
        {15pt}{0pt}{15pt}%
        {4pt}{0pt}%
```

```
            {0pt}{15pt}%
}
```

Level 4.

```
\def\@BeginItemizeLevelFour{%
    \BeginAList
        {15pt}{0pt}{15pt}%
        {4pt}{0pt}%
        {0pt}{15pt}%
}
\catcode`\@ = 12
```

• End of ts-itize.tip •

11.3.5 An Example Application of the Preceding Macros

You already have seen numerous examples of enumerated lists in this series. Here is the input to an example, because you should see the proper input format at least once. In particular note the indentation of the source code to improve readability on the screen.

```
\BeginEnumerate
    \Item This is great fun, I think. I hope you think so too.
        Again this paragraph is one of those useful and useless
        ones, because we need text and text and text.

        Now this paragraph is kind of fun! This paragraph is
        unlabeled!
    \Item More here. And then remember: we talked about nested
        lists.
        \BeginEnumerate
            \Item Have
            \Item Fun
            \Item You
            \Item Goofballs
            \Item As much
            \Item Fun
            \Item As
            \Item You
            \Item Can
            \Item Stand
        \EndEnumerate
\EndEnumerate
```

The output generated by the preceding source code reads as follows:

1. This is great fun, I think. I hope you think so too. Again this paragraph is one of those useful and useless ones, because we need text and text and text.

   Now this paragraph is kind of fun! This paragraph is unlabeled!
2. More here. And then remember: we talked about nested lists.

(a) Have
(b) Fun
(c) You
(d) Goofballs
(e) As much
(f) Fun
(g) As
(h) You
(i) Can
(j) Stand

11.4 Hanging Indentations (**\hangindent**, **\hangafter**)

Four different paragraph shapes based on hanging indentation can be generated in TEX. A paragraph with hanging indentation is generated by first assigning a value to the following two registers. The assignments to these two registers are followed by the text of the paragraph itself (with no empty line intervening).

1. A number must be assigned to TEX's integer parameter **\hangafter**.
2. A dimension value must be assigned to TEX's dimension parameter **\hangindent**.

The four different paragraph shapes which can be generated by assigning the various paragraph shapes are summarized in Fig. 11.4, p. 82.

Note that if you assign only a value to **\hangindent** but not to **\hangafter** an implicit assignment of a value of 1 to **\hangafter** is assumed.

11.4.1 The Counter Parameter **\hangafter**

The value of the counter parameter **\hangafter** (n in the figure mentioned above) specifies *to which lines the hanging indentation is applied.* This value is interpreted as follows:

1. If the value of this register is *negative*, then the indentation is applied to the first $-n$ lines.
2. If this counter register is *positive*, the hanging indentation is *not applied* to the first n lines; instead the indentation starts with line $n+1$ and continues until the paragraph of the paragraph.

11.4.2 The Dimension Parameter \hangindent

The dimension register **\hangindent** specifies the *amount of indentation.*

1. If the dimension assigned to this dimension register is *positive*, the indentation occurs on the *left*-hand side.
2. If the dimension assigned to this dimension register is *negative*, the indentation occurs on the *right*-hand side.

If a paragraphs' shape shape is controlled by setting **\hangindent** and **\hangafter**, then the values of **\leftskip**, **\rightskip**, **\parindent**, etc., are still taken into account. For instance, in order to generate the four examples in Fig. 11.4 on the next page where each of the four paragraphs is not indented it had to be either preceded by a **\noindent**, or the paragraph indentation has to be set to zero by **\parindent = 0pt**. Otherwise the first lines of all paragraphs would appear indented.

\hangindent and **\hangafter** can be set anywhere in the paragraph, and can be changed in the middle of a paragraph. The values for **\hangindent** and **\hangafter** as they are in effect *when the paragraph ends* are the values which are applied to the whole paragraph. While it is more natural to enter the settings of **\hangindent** and **\hangafter** at the beginning of a paragraph, this is by no means necessary.

11.4.3 \hangindent and \hangafter Must Be Reloaded for Each Paragraph

Note that once **\hangindent** and **\hangafter** are set, their values apply *only* to the current paragraph (this is different from values assigned to other parameters as, for instance, **\leftskip** and **\rightskip**; in those cases these values stay in effect until changed again, for instance, by ending a group in which they were enclosed). **\hangindent** and **\hangafter** stay only in effect for the current paragraph and therefore have to be reloaded each time a new paragraph with hanging indentation is started.

The paragraph to which hanging indentation is applied must follow the assignments to **\hangindent** and **\hangafter** immediately, meaning without any empty line or **\par** between.

11.4.4 Another Example of Hanging Indentation

The following example shows hanging indentation done in TeX. It also shows that an implicit **\hangafter = 1** is assumed, if a value is assigned to **\hangafter** but not to **\hangindent**.

case 1 `\hangafter` < 0, `\hangindent` > 0:

For this paragraph `\hangafter` has a value of −2, and `\hangindent` has a value of 40.0*pt*. For *negative* `\hangafter` ($n < 0$) indentation is applied to the *first* $|n|$ *lines*. For *positive* `\hangindent` ($x > 0$) the lines are indented at the *left*. As you can see the indentation in our cases is applied to the first two lines only. The remaining lines are unchanged in their length. We just write some more text here so the paragraph becomes longer and it is more easily to see the effect of setting those two TEX parameters. This should be enough text for the example.

case 2 `\hangafter` < 0, `\hangindent` < 0:

For this paragraph `\hangafter` has a value of −2, and `\hangindent` has a value of −40.0*pt*. For *negative* `\hangafter` ($n < 0$) indentation is applied to the *first* $|n|$ *lines*—in this respect this example is identical to the previous case. For *negative* `\hangindent` ($x < 0$) the lines are now indented at the *right*. As you can see the indentation in our cases is applied to the first two lines only. The remaining lines are unchanged in their length. We just write some more text here so the paragraph becomes longer and it is more easily to see the effect of setting those two TEX parameters. This should be enough text for the example.

case 3 `\hangafter` > 0, `\hangindent` > 0:

For this paragraph `\hangafter` has a value of 2, and `\hangindent` has a value of 40.0*pt*. For *positive* `\hangafter` ($n > 0$) indentation is *not* applied to the *first* n lines, but is applied to lines $n+1, n+2, \ldots$. In our case we have an indentation of all lines below and including line 3. Because the value x of the indentation is *positive* the indentation is applied to the *left* side of the lines. We just write some more text here so the paragraph becomes longer and it is more easily to see the effect of setting those two TEX parameters. This should be enough text for the example.

case 4 `\hangafter` > 0, `\hangindent` < 0:

For this paragraph `\hangafter` has a value of 2, and `\hangindent` has a value of −40.0*pt*. For *positive* `\hangafter` ($n > 0$) indentation is *not* applied to the *first n lines*, but to lines $n+1, n+2, \ldots$. In our case we have an indentation of all lines below and including line 3. Because the value x of the indentation is *negative* the indentation is applied to the *right* side of the lines. We just write some more text here so the paragraph becomes longer and it is more easily to see the effect of setting those two TEX parameters. This should be enough text for the example.

Figure 11.4. Illustration of `\hangafter` and `\hangindent`.

For an example see Figure 11.5 on the next page. The source code used to generate this figure follows below.

```
$$
    \vbox{
        \hsize = 4.0in
        \parskip = 12pt plus 2pt minus 1pt
        \parindent = 0pt

        \hangindent = 40pt
        \hangafter  = -2
        \noindent
            {\tt [1]} This is some hanging indentation stuff to show
        that those values have to be reloaded each time a new
        paragraph is started. So here we have a paragraph where we
        have hanging indentation.

            {\tt [2]} Here is another one, but this time we have no
        indentation. This is really fun, yes really, really,
        really, really, really, really, really, really.

        \hangindent = 40pt
        \hangafter  = -2

        {\tt [3]} And here we made the mistake of leaving an empty
        line before the paragraph. Oh well, that means that the
        hanging indentation will also not be applied to this
        paragraph.

        \hangindent = 40pt
        \hangafter  = -2
        \noindent
        {\tt [4]} Now it's right again. Now it's right again. Now
        it's right again. Now it's right again. Now it's right
        again. Now it's right again. Now it's right again. Now
        it's right again. Now it's right again. Now it's right
        again. Now it's right again. Now it's right again.

        \hangindent = 40pt
        \hangafter  = -2
        {\tt [5]} This is indented, I mean this paragraph because we
        did not write a {\tt\string\noindent}.  Now it's right
        again. Now it's right again. Now it's right again. Now it's
        right again. Now it's right again. Now it's right again. Now
        it's right again. Now it's right again. Now it's right again.
        Now it's right again. Now it's right again.
    }
$$
```

[1] This is some hanging indentation stuff to show that those values have to be reloaded each time a new paragraph is started. So here we have a paragraph where we have hanging indentation.

[2] Here is another one, but this time we have no indentation. This is really fun, yes really, really, really, really, really, really, really, really.

[3] And here we made the mistake of leaving an empty line before the paragraph. Oh well, that means that the hanging indentation will also not be applied to this paragraph.

[4] Now it's right again. Now it's right again. Now it's right again. Now it's right again. Now it's right again. Now it's right again. Now it's right again. Now it's right again. Now it's right again. Now it's right again. Now it's right again. Now it's right again. Now it's right again.

[5] This is indented, I mean this paragraph because we did not write a `\noindent`. Now it's right again. Now it's right again. Now it's right again. Now it's right again. Now it's right again. Now it's right again. Now it's right again. Now it's right again. Now it's right again. Now it's right again. Now it's right again. Now it's right again. Now it's right again.

Figure 11.5. Showing hanging indentation again.

11.4.5 Use of `\hangafter` and `\hangindent` to Generate Lists

I will now show an application of `\hangafter` and `\hangindent`, which shows an alternative way to generate lists—compare the following solution with the ideas presented in 10.6.2, p. 15.

In case an indented paragraph similar to the paragraphs of 10.6.2, p. 15, is needed (with no indentation on the right-hand side) such a paragraph shape can be generated following case 3 of Fig. 11.4, p. 82. By setting `\hangafter` to 1, TeX is instructed not to apply the hanging indentation to the first line, which means the hanging indentation applies to all *but* the first line. The first line of the paragraph must therefore begin with an hbox which is as wide as the hanging indentation, and this hbox contains the label positioned properly within this hbox.

Let me present the following example now. The `\HboxR` macro is used to make the hbox which begins the paragraph clearly visible. The following source code was used:

```
\InputD{box-mac.tip}                          % 9.3.14, p. I-343.
\hangafter = 1
\hangindent = 100pt
\noindent
\HboxR to \hangindent{\hskip 25pt\it Something\hfil}%
    This is an indented paragraph. Do you see that it
    works as you wanted? Well, good luck to you.
```

This source code generates the following output:

Something This is an indented paragraph. Do you see that it works as you wanted? Well, good luck to you.

11.4.6 Starting a Paragraph with Big Letters, \BigLetPar

Now I present a macro, which allows you to use one or more tall letters at the beginning of a paragraph. This requires the text of the paragraph itself to wrap around this initial big letter text. The macro **\BigLetPar** defined here relies on hanging indentation to achieve this floating around effect.

The source code of **\BigLetPar** begins now.

$\mathcal{P}'$ • **bletpar.tip** •

```
\InputD{box-mac.tip}                          % 9.3.14, p. I-343.
\catcode`\@ = 11
```

The following dimension register will hold the amount by which the box containing the special text will be moved down.

```
\newdimen\@BigLetDown
```

We need to allocate two dimension registers. Also a box register is needed.

```
\newdimen\@BigLetDimen
\newcount\@BigLetCount
\newbox\@BigLetBox
```

The horizontal separation between the initial "big letter(s)" and the paragraph's text of the first couple of lines is defined here (user changeable).

```
\newdimen\BigLetSep
\BigLetSep = 2pt
```

Define the minimum separation between the baseline of the big letter text and the top of the first line of text of the paragraph which appears under the big letter text (user changeable).

```
\newdimen\BigLetH
\BigLetH = 2pt
```

Define macro **\BigLetPar** now. This macro has one parameter, **#1**, the initial big letter text. This parameter should include the font change to the font to be used for the big letter text. A call to this macro should be followed immediately by the text of the paragraph itself. There must be no empty line between the call to this macro and the following text. This macro computes the hanging indentation parameters and then prints the big letter text with the rest of the text floating around the big letter text.

```
\def\BigLetPar #1{%
```

Finish the previous paragraph, just in case TEX has not yet done so.

```
        \par
```

Store big letter text in a box. Make it slightly wider so it later does not touch the text of the paragraph.

```
        \setbox\@BigLetBox = \hbox{#1\hskip\BigLetSep}%
```

Measure how far the big letter text has to be moved down. Then move it down (the big letter text moved down is stored in box `\@BigLetBox`).

```
        \setbox\@BigLetBox = \vtop{%
            \dimen0 = \baselineskip
            \offinterlineskip
```

The following empty hbox sits on the base line of the box being constructed here.

```
            \hbox{}
```

The top of the special text starts with the top of the regular text of this paragraph and therefore the current baseline has to be moved up a little bit.

```
            \vskip -0.7\dimen0
            \box\@BigLetBox
            \vbox to \BigLetH{}
        }%
```

Compute the number of lines to which hanging indentation must be applied.

```
        \@BigLetDimen = \dp\@BigLetBox
        \advance\@BigLetDimen by 0.7\baselineskip
        \advance\@BigLetDimen by 1.0\baselineskip
        \divide\@BigLetDimen by \baselineskip
        \@BigLetCount = \@BigLetDimen
```

Ignore the depth of the big letter box, because the depth of the big letter box should not influence the spacing within the paragraph itself.

```
        \dp\@BigLetBox = 0pt
```

Set-up the hanging indentation.

```
        \hangafter = -\@BigLetCount
        \hangindent = \wd\@BigLetBox
```

Start the paragraph and suppress the paragraph indentation.

```
        \noindent
```

Typeset the big letter text which is followed by the regular text.

```
        \hskip -\hangindent
        \box \@BigLetBox
```

Ignore any spaces following this macro's argument.

```
        \ignorespaces
}
\catcode`\@ = 12
```

• End of bletpar.tip •

Here is an example application of this macro. The following source code was used:

```
\InputD{bletpar.tip}                              % 11.4.6, p. 85.
\font\hugebf = cmbx12 scaled \magstep 3
$$
    \vbox{
        \hsize = 4.5in
        \BigLetPar{\hugebf A}
        nd here is an application so you
        can see how this macro works and what it does. Good luck
        for your own applications of this macro. Well we have
        to write a little more text
        so you can appreciate this macro fully.

        \BigLetPar{\hugebf The}
        text which is printed in big
        letters can be actually more than one letter as this
        example shows. The font size we choose causes this macro
        to use two lines for the text itself.
    }
$$
```

This source code generates the following output:

A nd here is an application so you can see how this macro works and what it does. Good luck for your own applications of this macro. Well we have to write a little more text so you can appreciate this macro fully.

The text which is printed in big letters can be actually more than one letter as this example shows. The font size we choose causes this macro to use two lines for the text itself.

11.4.7 Plain Format Macros \item and \itemitem

The plain TeX macros **\item** and **\itemitem** are based on **\hangindent** and **\hangafter**. Let me give their definitions here (slightly modified for better readability).

These macros do *not* contain instructions **\hangafter = 1**, because this instruction is redundant (see 11.4.3, p. 81). These macros generate a paragraph shape as it is shown in case 3 of Fig. 11.4, p. 82, with $n = 1$. Observe that the indentation is *not* applied to the first line. But the first line of each of the paragraphs is indented by using **\indent**.

`\indent` is also necessary because the paragraph starts with an hbox generating macro `\llap` and therefore TEX must be instructed to enter horizontal mode; see 6.5.2, p. I-182.

The macros are built in such a way that what seems to be an argument of `\item` or `\itemitem` really is an argument of `\textindent`. This is discussed in 22.3, p. III-218.

Here are the plain format definitions of `\item` and `\itemitem`.

$\mathcal{P}'$ • **itemplan.tip** •

```
\def\hang{%
    \hangindent = \parindent
}
\def\item{%
    \par
    \hang
    \textindent
}
\def\itemitem{%
    \par
    \indent
    \hangindent = 2\parindent
    \textindent
}
```

The macro `\textindent` is called after the hanging indentation is set up. Note that this macro has as argument an item label, which it prints. The item labels generated are right flush (simply go from the definition of `\llap` and compare the application of `\llap` below with the definition of `\LabelLR`).

```
\def\textindent #1{%
    \indent
    \llap{#1\enspace}%
    \ignorespaces
}
```

I found the addition of the following macro `\itemitemitem` useful (for an application that I cannot remember right now). The macro obviously generates a paragraph with three times the paragraph indentation as indentation of the whole paragraph.

```
\def\itemitemitem{%
    \par
    \indent
    \indent
    \hangindent = 3\parindent
    \textindent
}
```

• End of `itemplan.tip` •

Here is a sample application of the preceding macros. The following input was used. If you use `\itemitemitem`, you need to load the preceding source code file, because all other macro definitions are part of the plain format.

```
\input inputd.tip
\InputD{itemplan.tip}                       % 11.4.7, p. 88.

\parindent = 30pt
\parskip = 12pt plus 2pt minus 1pt

    This is some regular text which precedes the {\tt\string\Item}
examples. So let's get started now.
\item{1.} First item.
\item{2.} Second item.

    Here is some explanation of the second item. Well, that could
    be longer.  Here is some explanation of the second item. Well,
    that could be longer. Here is some explanation of the second
    item.

\item{} Here is some explanation of the second item. Well, that
    could be longer. Here is some explanation of the second item.
    Well, that could be longer. Here is some explanation of
    the second item.
\item{10.} Tenth item, with some subitems.
    \itemitem{(a)} Subitem a of \#10.
    \itemitem{(b)} Subitem b of \#10.
    \itemitem{(c)} Subitem c of \#10.

That's it for the example.
```

The preceding source code generates the following output:

This is some regular text which precedes the `\Item` examples. So let's get started now.

1. First item.

2. Second item.

Here is some explanation of the second item. Well, that could be longer. Here is some explanation of the second item. Well, that could be longer. Here is some explanation of the second item.

Here is some explanation of the second item. Well, that could be longer. Here is some explanation of the second item. Well, that could be longer. Here is some explanation of the second item.

10. Tenth item, with some subitems.

    (a) Subitem a of #10.

    (b) Subitem b of #10.

(c) Subitem c of #10.

That's it for the example.

11.5 A Generic Heading Printing Macro

A *generic heading printing macro* is used to generate section headings and such, for instance, for this series. It is called generic, because it is not only used for sections but also subsections, subsubsections and so forth, and it can be customized in many ways. It is *not* used for chapter headings in this series, because chapter headings are laid out quite differently.

Note that a user will normally *not* call the following macro directly. Instead the format being used, or the style files of the format being used, defines macros such as `\Section` or `\SubSection`. These macros in turn call `\GenericHeading`.

Note also that the following macro does *not* increment the section, subsection or whatever number needs to be incremented. This function must be implemented inside `\Section`, `\SubSection`, and so forth.

11.5.1 The Parameters of `\GenericHeading`

The macro `\GenericHeading` has 12 parameters (read on for how to find that this is possible). Note that reading the description of all the parameters below does not tell you the full story: you have to read the source code of this macro which follows later to find out about many additional details and other assumptions of this macro.

Note that a heading is subdivided into the *number* of the heading (such as 12.3.4) and the *heading text* itself (such as "An Example Application"). The macro accommodates headings without number.

The macro below is listed as having 12 parameters. There is no such thing as a macro with 12 parameters in TEX, but a trick can be used to simulate a macro with more than 9 parameters. See 22.3.2, p. III-219, for details.

- `#1`. The level of the entry, for instance, 0 for the part of a document, 1 for the chapter of a document, 2 for a section of a document, 3 for a subsection of a document and so forth. This information is currently not used by `\GenericHeading`.
- `#2`. The vertical space (a vertical glue value) to be generated *before* the heading itself. The `\MaxVskip` macro is used (see 8.6, p. I-309), so the amount of vertical skip is the maximum of the amount of vertical glue specified with this parameter and the value of glue inserted after the preceding text.

This is a natural thing to do. Assume, for instance, that the text preceding a heading is one of the lists which were discussed before. Then the amount of glue before the heading should *not* be the sum of the glue inserted at the end of the preceding list and this parameter, but the maximum of the two values.

- `#3`. Specify a value of 1 for this parameter if you want ragged right headings. 0 gives you right flush headings.
- `#4`. Specify a value of 1 here, if hyphenation is *not* acceptable in a heading. 0 means that hyphenation is acceptable.
- `#5`. This parameter controls the indentation of the second, third, and so forth line, *if* the heading is longer than one line. Note: a heading is printed together with its associated number and the question is where the left margin of the second (and further) lines fall, left flush with the number (`#5` must be 1 in this case) or left flush with the heading's text (`#5` must be 0 in this case).
- `#6`. The vertical space (a vertical glue value) inserted *after* the heading (that is usually the distance between the heading and the first line of the paragraph following the heading).
- `#7`. This parameter controls the indentation of the first paragraph following the heading. Specify 0, if the paragraph following this heading should *not* be indented. Specify as 1 if the first paragraph should be indented.

  Note that if the paragraph indentation of a document is zero anyway, the value of this parameter does not matter.
- `#8`. The font change instruction to be used to generate the heading such as `\Large\bf`. Note that this instruction is assumed to also change `\baselineskip` appropriately for printing a multiline heading, if the heading should be longer than one line.
- `#9`. This parameter is a dimension which is interpreted differently depending on whether it is positive or negative.

  - If this parameter is *positive*, then this parameter specifies the horizontal space the heading's number is allowed to occupy. The heading will be positioned left flush within this space.

    In case the text of the heading exceeds one line, the indentation of the lines after the first one are controlled by `#5`.
  - If this parameter is *negative*, then this dimension is first converted into a positive dimension by taking the absolute value. This value is then interpreted as the horizontal space to be inserted *between* the heading's number and the heading's text itself.

    In case the heading's text wraps around, that is it longer than one line, then the indentation of those lines is determined by `#5`.

- `#10`. The number of the heading (note that the number of the section is treated separately from the text of the heading itself). The number of the heading is printed left flush.
- `#11`. The text of the heading itself.
- `#12`. The text of the heading as it will appear in the table of contents (usually this and the preceding parameter are identical). The following Subsection discusses possible reasons for having separate parameters.

11.5.2 Token Register `\EveryHeading`

The token register `\EveryHeading` which by default is empty is executed each time right after macro `\GenericHeading` was called. This allows one to execute some source code before the real processing of `\EveryHeading` begins.

11.5.3 Controlling Linebreaks

Note that there are two separate parameters, `#11` and `#12`, which seem to be identical all the time. Sometimes TEX does not find a proper line break in the heading printed or in the table of contents entry; in this case the line break must be done manually ("`\hfill\break`" must be inserted). Usually, when such a line break is inserted by hand, then this line break is either suitable for the heading but not for the table of contents, or vice versa. Therefore the macro here offers the possibility of a separate specification for heading and table of contents entry.

Usually things can be further simplified as follows: make `#11` and `#12` identical and insert a call of either macro `\LineBreakHeading` or `\LineBreakToc` in the heading's text. These macros are set up below in such a way that they expand properly. Thus `\LineBreakHeading` expands to `\hfil\break` when the heading is printed, whereas it expands simply to a space, when the table of contents information is written out. The macro `\LineBreakToc` works in the reverse to `\LineBreakHeading`. See further below for details.

Additionally below a macro `\IgnoreInRunningHead` is defined. This macro essentially has no effect (it has one argument `#1` which is also its expansion). This macro's definition though is changed to `$\ldots$` when the running head is printed (more precisely when the `\mark` instruction is executed which later causes the running head to be printed). Calling this macro allows one to accommodate long section titles (longer than the running head). See 31.2.9, p. III-605, for that redefinition of `\IgnoreInRunningHead`.

11.5.4 Running Heads

I should note that the title of a *section* is actually used for *three* purposes: in addition for the printing of the title on a page and the entry into the table of contents it is also used as running head (so is the chapter title). If the title is too long in the running head, then the current macro does not allow for correction (there is no second optional argument). Such an extension would be fairly easy to add to the macro discussed here (I "cheated" and kept titles of sections short enough to cause no problems).

11.5.5 The Macro Source Code

Here is the source code of the generic heading macro. This source code assumes that you handle the paragraph indentation according to 10.9.4, p. 35.

$\mathcal{P}'$ • **genhead.tip** •

```
\InputD{vsmax.tip}                        % 8.6, p. I-310.
\InputD{box-mac.tip}                      % 9.3.14, p. I-343.
\InputD{parin.tip}                        % 10.9.4, p. 35.
```

The token register **\EveryHeading** is allocated here.

```
\newtoks\EveryHeading
\EveryHeading = {}
\catcode`\@ = 11
```

Two counter registers to save **#1** and **#5**.

```
\newcount\@GenericHeadingCount
\newcount\@GenericHeadingIndent
```

Define macros **\LineBreakHeading** and **\LineBreakToc** to generate an error message because they may not appear outside a heading. These two macros are redefined twice below. Also note that yet different definitions apply to them when a running head it typeset; see 31.2.9, p. III-605, for details.

```
\def\LineBreakHeading{%
    \errmessage{%
        \string\LineBreakHeading/\string\LineBreakToc:
            can only be used inside a heading.}%
}
\let\LineBreakToc = \LineBreakHeading
\def\IgnoreInRunnningHead #1{%
    #1%
}
```

Now define the macro **\GenericHeading**.

```
\def\GenericHeading #1#2#3#4#5{%
```

Just to be sure issue a **\par** so that if the preceding paragraph has not been terminated it is now.

```
    \par
    \the\EveryHeading
    \@GenericHeadingCount = #1
    \@GenericHeadingIndent = #5
```

Then generate the vertical space between the preceding paragraph and the heading text. Note that the vertical space to be generated should be at least **#1**, and any vertical glue which is already on the main vertical list should be taken into account. That is the reason why macro **\MaxVskip** is used (rather than simply inserting **\vskip #2**).

```
    \MaxVskip{#2}%
```

I found to test whether `\leftskip` is zero (and to print a message if it is not) to be very useful. The most common reason for setting `\leftskip` to a non-zero value is to implement a list, and lists by definition don't go beyond sections, subsections and so forth. Therefore, this macro prints a warning, if `\leftskip` is non-zero.

```
        \ifdim\leftskip = 0pt
        \else
            \message{\string\GenericHeading: \noexpand\leftskip
                is non-zero, forgotten to terminate a list?}%
        \fi
```

Now it is time to generate the heading itself. Note that we try to avoid a page break within the heading (`\interlinepenalty = 10000` does that; see 32.5.4.2, item 1.d, p. IV-21, for details). The heading itself is typeset as a paragraph. All changes below are done in a group. Note that the vertical spacing was handled using `\MaxVskip`, and therefore `\parskip` is set to zero.

```
        \begingroup
        \interlinepenalty = 10000
        \parindent = 0pt
        \parskip = 0pt
```

Prevent hyphenation, if requested.

```
        \ifnum #4 = 1
            \hyphenpenalty = 10000
        \fi
```

Use ragged right if requested.

```
        \ifnum #3 = 1
            \rightskip = 0pt plus 50pt
        \fi
```

Now continue execution by calling macro `\@GenericHeading`.

```
        \@GenericHeading
}
```

What is defined as parameter #i for `\GenericHeading` is parameter #$i - 5$ of `\@GenericHeading`.

```
\def\@GenericHeading #1#2#3#4#5#6#7{%
```

Do the font change for this heading now.

```
            #3%
```

The macro `\LineBreakHeading` should generate a line break, but `\LineBreakToc` should not, when the heading itself is printed in the text (which is currently the case).

```
            \def\LineBreakHeading{\hfil\break}%
            \def\LineBreakToc{ }%
```

There are two different formats concerning the relationship between the heading's number and the heading's text. First deal with `#6` positive.

```
        \ifdim #4 > 0pt
```

Does the text need to be indented if it wraps around (#5 of `\GenericHeading`)?

```
            \ifnum\@GenericHeadingIndent = 0
```

Yes, it needs to be indented, making it flush with the heading's text. The macro uses hanging indentation explained in 11.4, p. 80.

```
                \hangindent = #4
                \hangafter = 1
                \leavevmode
                \hbox to #4{#5\hfil}%
            \else
```

No indentation if text wraps around, that is flush with the heading's number.

```
                \leavevmode
                \hbox to #4{#5\hfil}%
            \fi
        \else
```

If #9 of `\GenericHeading` is negative, then $-\#9$ specifies the amount of space between the heading's number and the heading's text.

```
            \setbox0 = \hbox{#5\hskip -#4}
            \ifnum\@GenericHeadingIndent = 0
                \hangindent = \wd0
                \hangafter = 1
                \leavevmode
                \box0
            \else
                \leavevmode
                \box0
            \fi
        \fi
```

Write the heading's text.

```
        #6%
```

Write to the table of contents file, if such a mechanism is set up in the first place. Writing the table of contents file is done by calling `\WriteToAuxSpecial`; see 30.7.5.1, p. III-555, for details.

```
        \if\NameDefinedConditional{WriteToAuxSpecial}%
```

The definitions of the `\LineBreak...` macros must be the inverse of what the definitions were before, because the table of contents entry is being written now.

```
            \def\LineBreakToc{\hfil\break}%
            \def\LineBreakHeading{ }%
            \WriteToAuxSpecial{toc}{\the\@GenericHeadingCount}%
                {#5}{#7}{\PrintCounter{PageNo}}%
        \fi
```

Finish the heading that was typeset as a paragraph. The heading was not terminated earlier so that the table of contents writing instruction is part of the heading's paragraph and therefore the page number printed in the table of contents is correct; see 28.5.5, p. III-475, for details.

```
          \par
```

Terminate the group in which the typesetting of the heading was enclosed.

```
      \endgroup
```

Now generate the vertical space between the heading and the first line of the following text. Prevent a page break between these two items.

```
      \nobreak
      \vskip #1
```

Suppress the paragraph indentation of the first paragraph following this heading, if #7 of **\GenericHeading** is zero. The technique used here is explained in 10.9.4, p. 35.

```
      \ifnum #2 = 0
          \SuppressNextParIndent
      \fi
}
\catcode`\@ = 12
```

• End of **genhead.tip** •

11.5.6 An Example Using the Preceding Macro

Let me present now an example using the preceding heading macro. This example is structured in a way the macro **\GenericHeading** would normally be used: the user does *not* directly invoke this macro, but this macro is called by other macros such as **\SectionX**, **\SubSectionX** and **\SubSubSectionX**.

Let's be elegant and store the definitions of these macros in a style file. This style file contains font definitions and macro definitions.

• **ex-genhead-sty.tip** •

```
\InputD{genhead.tip}                      % 11.5.5, p. 93.
```

Get some large boldface fonts for the headings and define some macros switching to those fonts and changing **\baselineskip** at the same time.

```
\font\Bf = cmbx12
\font\BBf = cmbx12 scaled \magstep 1

\def\BfChange {\Bf  \baselineskip = 14pt}
\def\BBfChange{\BBf \baselineskip = 20pt}
```

Next define macros **\SectionX**, **\SubSectionX** and **\SubSubSectionX**. All macros have two parameters. The first parameter, **#1**, is the number of the heading being generated, the second parameter, **#2**, is the text of the heading. Note that these macro definitions would normally be part of the format you use or stored in a style file of the format you use.

The following definitions are atypical in that they do not provide for any stretchability or shrinkability of the glue around headings, which is atypical for a bottom flush page layout. Compare the settings below with the settings of 31.2.9, p. III-604. Also, I used special names for these macros ending in `X` to avoid conflicts with the names of the heading macros defined for this series. See the same reference for details.

Parameter `#6` controls the relationship of the heading's number and the heading's text and also controls the positioning of any continuation lines of the heading's text. Note that the setting of `#6` is inconsistent if you compare the settings in the first two macros with the setting in the last macro. This was done on purpose, to expose these two different styles.

```
\def\SectionX #1#2{%
    \GenericHeading{2}{20pt}{0}{0}{0}{10pt}{0}%
        {\BBfChange}{72pt}{#1}{#2}{#2}%
}
\def\SubSectionX #1#2{%
    \GenericHeading{3}{15pt}{0}{0}{0}{5pt}{0}%
        {\BfChange}{40pt}{#1}{#2}{#2}%
}
\def\SubSubSectionX #1#2{%
    \GenericHeading{4}{5pt}{0}{0}{1}{0pt}{0}%
        {\bf}{-35pt}{#1}{#2}{#2}%
}
```

• End of `ex-genhead-sty.tip` •

After the definition of this "pseudo style file" we have the source code of the example.

• `ex-genhead.tip` •

```
\input inputd.tip
```

Load the style file just defined.

```
\InputD{ex-genhead-sty.tip}                  % 11.5.6, p. 96.
```

We need to disable writing table of contents information, because the following information is not supposed to be printed as part of the table of contents of this volume (but the macros to write to the table of contents are already loaded). `\WriteToAuxSpecial` is disabled to achieve this effect. Because of this redefinition this example must be later executed in a group.

```
\let\WriteToAuxSpecial = \GobbleFive
```

Now use the previous macros.

```
\SectionX{1.2}{This Is Fun and Let Me Make It a Little
    Longer Now, So It Wraps Around and So On}
Ok and here is some text. Ok and here is some text.
Ok and here is some text. Ok and here is some text.

Ok and here is some text. Ok and here is some text.
Ok and here is some text. Ok and here is some text.
Ok and here is some text. Ok and here is some text.
```

```
\SubSectionX{1.2.1}{This Is Fun and Let Me Make It
     Also a Little Longer Now So It Wraps Around and So On}
Ok and here is some text. Ok and here is some text.
Ok and here is some text. Ok and here is some text.
Ok and here is some text. Ok and here is some text.

Ok and here is some text. Ok and here is some text.
Ok and here is some text. Ok and here is some text.
Ok and here is some text. Ok and here is some text.

Ok and here is some text. Ok and here is some text.

\SubSectionX{1.2.2}{This Is Fun, and Short}
\SubSubSectionX{1.2.2.1}{This is Fun and Short}
     Let me write some text here. Oh well, doesn't it work
     very nicely?!
\SubSubSectionX{1.2.2.2}{This Is Fun and Let Me Make It Longer
     Now So It Wraps Around and So On}
\SubSubSectionX{1.2.2.3}{This Is Fun and Let Me Make It Longer
     and Longer and Longer, So It wraps Around and So On.
     This Also Shows the Different Style in the Selection of
     the Wrap Around Modus.}
     And here is some more text after the preceding subsubsection.
\bye
```

• End of `ex-genhead.tip` •

Now let me print the output generated by the preceding TEX code:

1.2 This Is Fun and Let Me Make It a Little Longer Now, So It Wraps Around and So On

Ok and here is some text. Ok and here is some text. Ok and here is some text. Ok and here is some text.

Ok and here is some text. Ok and here is some text. Ok and here is some text. Ok and here is some text. Ok and here is some text. Ok and here is some text.

1.2.1 This Is Fun and Let Me Make It Also a Little Longer Now So It Wraps Around and So On

Ok and here is some text. Ok and here is some text. Ok and here is some text. Ok and here is some text. Ok and here is some text. Ok and here is some text.

Ok and here is some text. Ok and here is some text. Ok and here is some text. Ok and here is some text. Ok and here is some text. Ok and here is some text.

Ok and here is some text. Ok and here is some text.

1.2.2 This Is Fun, and Short

1.2.2.1 This is Fun and Short
Let me write some text here. Oh well, doesn't it work very nicely?!

1.2.2.2 This Is Fun and Let Me Make It Longer Now So It Wraps Around and So On

1.2.2.3 This Is Fun and Let Me Make It Longer and Longer and Longer, So It wraps Around and So On. This Also Shows the Different Style in the Selection of the Wrap Around Modus.
And here is some more text after the preceding subsubsection.

11.6 Using `\leftskip` and `\rightskip` with Stretchability

So far we have used `\leftskip` and `\rightskip` always as glues without any stretchability. Let us see now what can be done when we allow these glues to stretch (and shrink).

11.6.1 Ragged Right and Ragged Left Text, `\raggedright`

Let me discuss how ragged right lines are generated in TEX. When this type of output is used with the typewriter font, then TEX's output looks as if it came straight from a typewriter. But also with proportionally spaced fonts one may choose ragged right type of output, in particular for very narrow columns.

A ragged right layout can be achieved with a proper setting of `\rightskip` by giving this glue some stretchability so lines are allowed to *end* at different horizontal locations.

The macro `\raggedright` macro of the plain format is defined as follows:

```
\def\raggedright{%
     \rightskip = 0pt plus 2em
     \spaceskip = .3333em
     \xspaceskip = .5em
}
```

Note that in addition to giving `\rightskip` some stretchability, `\spaceskip` and `\xspaceskip` are set to non-zero values and are defined to be glues with *no* stretchability and shrinkability. By default in proportionally spaced fonts interword glue is allowed to stretch and shrink to allow for left and right justification, but this type of interword glue is only in effect if `\spaceskip` and `\xspaceskip` are both zero. The interword glue should *not* be allowed to stretch or shrink under those circumstances which is achieved by setting `\spaceskip` and `\xspaceskip` to non-stretchable and non-shrinkable glues; see 16.2.4, p. 277. The only glue which can be stretched to adjust the line length is `\rightskip` now.

Here is the example of a paragraph with some ragged right text. The following input is used in this example (note that `\leftskip` and `\rightskip` are set to values which result in the left and right margins being moved in on both sides):

```
\rightskip = 1.5in plus 4em
\leftskip =  1.5in
\spaceskip = .3333em
\xspaceskip = .5em

This is some sample text for ragged  right.
We don't use the {\tt\string\raggedright} macro
because we use an indentation of the margins on both
sides. This text should suffice to show how the output
comes out. That's it for now.
```

This input generated the following output:

This is some sample text for ragged right. We don't use the `\raggedright` macro because we use an indentation of the margins on both sides. This text should suffice to show how the output comes out. That's it for now.

The same idea can be applied to generate ragged left text by specifying `\leftskip` as glue which can stretch. This is less common than ragged right text though and therefore no macro is introduced here.

11.6.2 Automatically Centered Text

Using the ideas of stretchable `\rightskip` and `\leftskip` glue one can center multiple lines of text. The centering of single lines can be achieved by using `\centerline` (see 6.6.1, p. I-185), in which case the user must make the decisions about line breaks. We are looking for a different approach in which TeX determines line breaks automatically, filling each line up as much as possible.

Here is an example showing how easily it can be done. The interword glue is now allowed to stretch, of course. Also `\parfillskip`, the glue inserted at the right-hand side of the last line of a paragraph, it is set to zero, because the last line is not supposed to be left flush but centered; see 10.11, p. 45.

```
\rightskip = 1in plus 4em
\leftskip = \rightskip
\spaceskip = .3333em
\xspaceskip = .5em
\parfillskip = 0pt
\noindent
On the other hand there are sometimes cases where you want to have a
sequence of lines, all centered, but you do not want to make the line
breaking decisions all by yourself. So here is a sample where \TeX{}
did all the work.
```

This input generates the following output:

On the other hand there are sometimes cases
where you want to have a sequence of lines,
all centered, but you do not want to make
the line breaking decisions all by yourself. So
here is a sample where TeX did all the work.

Let me reprint the same paragraph without `\parfillskip = 0pt` now so you can see the effect if you omitted this instruction: the last line is no longer centered but ragged right.

```
\rightskip = 3cm plus 4em
\leftskip = \rightskip
\spaceskip = .3333em
\xspaceskip = .5em
\parfillskip = 0pt plus 1fil        % Default!
\noindent
On the other hand there are sometimes cases where you want to have a
sequence of lines, all centered, but you do not want to make the line
breaking decisions all by yourself. So here is a sample where \TeX{}
did all the work.
```

This input generates the following output:

On the other hand there are sometimes
cases where you want to have a
sequence of lines, all centered, but you
do not want to make the line breaking
decisions all by yourself. So here is a
sample where TeX did all the work.

11.6.3 The Definition of `\BeginCenter` and `\EndCenter`

Based on the discussion in the preceding Subsection it is very easy to defined the two macros `\BeginCenter` and `\EndCenter`:

$\mathcal{P}'$ • `centerng.tip` •

```
\def\BeginCenter{%
```

Terminate any preceding paragraph.

```
      \par
```

Start a group.

```
      \begingroup
```

The following settings are derived from the preceding example.

```
      \rightskip = 1in plus 4em
      \leftskip = \rightskip
      \spaceskip = .3333em
      \xspaceskip = .5em
      \parfillskip = 0pt
      \noindent
}
```

Next is the definition of \EndCenter.

```
\def\EndCenter{%
```

Note: first terminate the paragraph (if the user did not do so) and then terminate the group started by \BeginCenter.

```
      \par
      \endgroup
}
```

• End of centerng.tip •

11.7 Summary

In this chapter we learned the following:

- Lists can be generated very easily in TeX. **\leftskip** and **\rightskip** are used to set the margins. Item labels (which are either right justified or left justified) are generated by zero width hboxes. Such hboxes are preceded with a **\leavevmode** call in order to trigger horizontal mode in TeX.
- Various macros to print such lists were discussed. The most complex macros of this kind as defined in this Chapter allow for the automatic enumeration of lists, and contain parameters for the proper vertical spacing. These macros are used to print all the lists in this series.
- The list macros of this chapter allow for the automatic numbering using digits, letters or other symbols. In the case of nested lists the form of enumeration is usually adjusted automatically.
- Paragraphs with hanging indentation can also be generated very easily in TeX using **\hangindent** and **\hangafter**.
- A generic header printing macro was discussed. This macro has numerous arguments which allow the user to generate almost any header. Also this

macro was used in the printing of this series. Usually a user will not call this macro directly, but will call other appropriately setup macros.

- Ragged right, ragged left or automatically centered text can be easily generated using `\leftskip` and `\rightskip`.

12
Paragraphs, Part III

This chapter continues the discussion of processing paragraphs in TeX. We will discuss **\parshape**, gives you an extremely detailed control over the shape of a paragraph. We will investigate timing issues in TeX, which are relevant when TeX switches to horizontal mode and back to vertical mode. We will furthermore investigate **\vadjust**, and discuss the typesetting of table of contents entry. In the end we have a detailed look at hyphenation and line break computations in general.

12.1 General Paragraph Shapes, \parshape

For the "ultimate" control to typeset a paragraph there is the TeX primitive **\parshape**. This instruction is applied as follows:

$$\texttt{\textbackslash parshape} = n \quad i_1\, l_1 \;\; i_2\, l_2 \;\; \ldots \;\; i_n\, l_n$$

where $n \geq 1$ is an integer, and all i_k and l_k ($1 \leq k \leq n$) are *dimensions.* The kth line of the paragraph following **\parshape** will have length l_k and this line will be indented from the left by i_k. If the resulting paragraph has *fewer* than n lines the additional (and superfluous) specifications are simply ignored. If the resulting paragraph has *more* than n lines, the specifications for the last line n are applied to all lines $k > n$.

12.1.1 A First \parshape Example

Let us start with a small **\parshape** example. The following input is used (line 7 in the text below refers to line 7 of the *output* produced by the following text):

```
\parindent = 0pt
\parshape = 7
      0.2in   4.0in   % 1
      0.4in   3.8in   % 2
```

```
    0.4in   4.0in   % 3
    0.8in   3.4in   % 4
    0.7in   3.0in   % 5
    0.6in   3.6in   % 6
    0.7in   3.8in   % 7
This shows an example of the {\tt\string\parshape} \TeX{} primitive.
We have not planned anything particular but just want to show the
application of this \TeX{} primitive. On page 101 of the \TeX{} book
you can find another interesting example. Observe that after line~7
the layout of line~7 is simply repeated ``for ever.'' This should
convince everybody that you can do everything with
{\tt\string\parshape}. On the other hand the {\tt\string\parshape}
primitive is rarely used, because this much
of a detailed control over \TeX's output is rarely needed.
```

The output from the previous source code looks as follows:

This shows an example of the **\parshape** TEX primitive. We have not planned anything particular but just want to show the application of this TEX primitive. On page 101 of the TEX book you can find another interesting example. Observe that after line 7 the layout of line 7 is simply repeated "for ever." This should convince everybody that you can do everything with **\parshape**. On the other hand the **\parshape** primitive is rarely used, because this much of a detailed control over TEX's output is rarely needed.

12.1.2 Another \parshape Example

Let us study another **\parshape** example. Here this command is used to generate a paragraph indented on both sides (previously we discussed an approach using **\leftskip** and **\rightskip** to achieve the same effect). If $n = 1$ in a **\parshape** command, then the specification l_1 and i_1 are applied to all lines of the paragraph.

Here is the input code for the following example:

```
\parshape = 1 1.0in 3.2in
   Here we show how we can use the {\tt\string\parshape} command
to print a paragraph with both sides indented. Beforehand
we achieved the same effect using {\tt\string\left\-skip} and
{\tt\string\rightskip}. Observe though, that the next paragraph
will no more be indented, unless we insert another
{\tt\string\parshape} command. That text should be sufficiently
long to demonstrate this use of the {\tt\string\par\-shape}
command.
```

The output generated by the preceding source code reads as follows:

Here we show how we can use the **\parshape** command to print a paragraph with both sides indented. Beforehand we achieved the same effect using

\leftskip and \rightskip. Observe though, that the next paragraph will no more be indented, unless we insert another \parshape command. That text should be sufficiently long to demonstrate this use of the \parshape command.

Note that \parshape still takes all other parameters into account. In the following \leftskip is set to −0.5 in which moves the left margin to the left, by 0.5 in. The \parshape command specified an indentation of *all* lines of 1.0 in. This means that all lines will be indented by 1.0 in + (−0.5 in), which is $0.5in$.

The following input is used for the example:

```
\leftskip = -0.5in
\parshape = 1 1.0in 3.2in
  Here we show how we can use the {\tt\string\parshape} command
to print a paragraph with both sides indented. Beforehand
we achieved the same effect using {\tt\string\leftskip} and
{\tt\string\rightskip}. Observe though, that the next paragraph
will no more be indented, unless we insert another
{\tt\string\parshape} command. The text should be sufficiently
long to demonstrate this use of the {\tt\string\parshape}
command.
```

The output generated by the preceding source code reads as follows:

Here we show how we can use the \parshape command to print a paragraph with both sides indented. Beforehand we achieved the same effect using \leftskip and \rightskip. Observe though, that the next paragraph will no more be indented, unless we insert another \parshape command. The text should be sufficiently long to demonstrate this use of the \parshape command.

12.1.3 An Extended \parshape Command, \XParShape

The macro \XParShape is an extended \parshape command and defined as follows:

$$\texttt{\textbackslash XParShape=}\ n \quad \begin{array}{lllll} m_1 & ind_1 & \delta_{ind_1} & len_1 & \delta_{len_1} \\ m_2 & ind_2 & \delta_{ind_2} & len_2 & \delta_{len_2} \\ \ldots & ind & & & \\ m_n & ind_n & \delta_{ind_n} & len_n & \delta_{len_n} \end{array}$$

The first parameter n is the number of the immediately following *line groups*. A line group j (there are n such line groups identified by an integer in the range $[1 \ldots n]$) applies to m_j lines, before the next line group $j+1$ starts to apply for the following m_{j+1} lines (lines refers to lines in the output, the same way it referred to lines in the output when the \parshape command was used).

Let me discuss one particular line group j now: the indentation of the kth line ($1 \leq k \leq m_j$) within this line group is $ind_j + (k-1) * \delta_{ind_j}$. The length of this line is computed as follows: $len_j + (k-1) * \delta_{len_j}$.

The main difference compared to the standard **\parshape** command is obviously that **\XParShape** groups lines into line groups. To those lines within their line groups, linear changes to the indentation and lengths are applied. The amount of these linear changes is part of the description in each line group.

The operations of the **\XParShape** macro is straightforward. It simply uses the above data to compute a regular **\parshape** command and then executes the computed **\parshape** command. In other words, it converts each line group j into m_j length and indentation pairs as the **\parshape** command is able to read it in and process it. Then **\parshape** is invoked.

As it is the case with **\parshape**, if too many lines are specified, the extra specifications are simply ignored and if too few lines are specified, the specification applicable to the very last line is repeated as often as necessary.

The source code for the definition of **\XParShape** begins now.

$\mathcal{P}'$ • **x-parsh.tip** •

```
\InputD{doloop.tip}                              % 27.1.8, p. III-412.
\catcode'\@ = 11
```

First three counters are declared: **\X@ParShapeCountA** counts the number of line groups (n). **\X@ParShapeCountB** is the loop variable. **\X@ParShapeCountC** counts how many lines the **\parshape** has (which is generated from within this macro).

```
\newcount\X@ParShapeCountA
\newcount\X@ParShapeCountB
\newcount\X@ParShapeCountC
```

Declare two dimension registers to hold intermediate values.

```
\newdimen\X@ParShapeDimenA
\newdimen\X@ParShapeDimenB
```

Now define the **\XParShape** command as explained before. This macro has no parameters. The value for n is assigned to **\X@ParShapeCountA** using **\afterassignment**. Then execution continues with **\X@ParShapeCountB**.

```
\def\XParShape{%
```

\X@ParShapeCollect must start out with a space. Collect all line groups in there.

```
      \def\X@ParShapeCollect{ }%
      \X@ParShapeCountC = 0
      \afterassignment\X@ParShapeB
      \X@ParShapeCountA
}
```

\X@ParShapeB is a macro to decide whether there is still a line group to be read in or not. If there is then this macro calls **\X@ParShapeC** to continue. If not then it calls **\X@ParShapeD** to stop.

```
\def\X@ParShapeB{%
```

```
    \ifnum\X@ParShapeCountA = 0
        \let\@XParShapeNext = \X@ParShapeD
    \else
```

Decrement counter and continue.

```
        \advance\X@ParShapeCountA by -1
        \let\@XParShapeNext = \X@ParShapeC
    \fi
```

Recursion, if at all, happens here.

```
    \@XParShapeNext
}
```

\X@ParShapeD is called in the end, after the **\parshape** command was built and therefore can be executed. **\X@ParShapeCountC** and **\X@ParShapeCollect** together expand to a complete **\parshape** command.

```
\def\X@ParShapeD{%
    \parshape = \X@ParShapeCountC\X@ParShapeCollect
}
```

The macro **\X@ParShapeC** absorbs one line group and adds the necessary dimension pairs to **\X@ParShapeCollect**. The macro has the following five parameters (all of them are delimited parameters, with the space being the delimiter):

- **#1**. How often to multiply (m_k).
- **#2**. Indentation (ind_k).
- **#3**. The delta of the indentation (δ_{ind_k}).
- **#4**. The line length (len_k).
- **#5**. The line length delta (δ_{len_k}).

```
\def\X@ParShapeC #1 #2 #3 #4 #5 {%
    \message{\string\X@ParShapeC: #1, #2, #3, #4, #5}%
    \X@ParShapeDimenA = #2%
    \X@ParShapeDimenB = #4%
    \DoLoop{\X@ParShapeCountB}{1}{1}{#1}%
        {%
            \edef\X@ParShapeCollect{%
                \space
                \X@ParShapeCollect
                \the\X@ParShapeDimenA
                \space
                \the\X@ParShapeDimenB
                \space
            }%
            \advance\X@ParShapeCountC by 1
            \advance\X@ParShapeDimenA by #3\relax
            \advance\X@ParShapeDimenB by #5\relax
        }%
```

The following macro calls will cause the next line group to be absorbed or the recursion to be stopped.

```
        \X@ParShapeB
}
\catcode`\@ = 12
```

• End of **x-parsh.tip** •

An example of using **\XParShape** can be found in Fig. 12.1 on the next page. The following source code was used to generated this figure (note that the first line of the generated paragraph is indented by **\parindent**, which is 20.0pt).

```
\input inputd.tip

\InputD{samplepa.tip}                    % 27.1.3.3, p. III-402.
\InputD{x-parsh.tip}                     % 12.1.3, p. 108.

\XParShape = 3
        10  70pt    0pt     200pt   5pt
        10  70pt    5pt     250pt   -5pt
        10  120pt   -5pt    200pt   10pt
        \SamplePar{X}{17}
```

12.2 Timing Issues in Paragraph Parameters

You have already seen that it is possible to change various parameters which determine the layout of a paragraph. Among the most important parameters are **\leftskip**, **\rightskip**, **\parskip** and **\baselineskip**.

Now we have to look into *changes to those parameters* in more detail. For instance, one question you might ask yourself is what happens if **\baselineskip** were changed in the middle of a paragraph? To which lines does the new value apply? To which lines does the old value apply?

For the following discussion two specific times in the processing of a paragraph are relevant:

1. One is the moment when TeX *begins* a paragraph and switches to horizontal mode.

   In 10.4, p. 11, we discussed when TeX starts a paragraph. In the following example either the first character starts a paragraph or **\leavevmode** does.
2. The other is the moment when TeX *ends* a paragraph.

   In 10.5, p. 12, we discussed how a paragraph is terminated. We will use \pars for our example to clearly mark the end of paragraphs.

There are two types of parameters:

1. For the *first set of parameters* the values of these parameters, as they are set at the *beginning of a paragraph*, are applied to the paragraph. These

Identification of this paragraph: *X. Sample paragraph 1, with 17 sentences.* So here we go, and when you check the number of sentences, then note that these first two sentences do *not* count. This is one of the many sentences this macro generates, to be more specific it is sentence number 1 of 17. This is one of the many sentences this macro generates, to be more specific it is sentence number 2 of 17. This is one of the many sentences this macro generates, to be more specific it is sentence number 3 of 17. This is one of the many sentences this macro generates, to be more specific it is sentence number 4 of 17. This is one of the many sentences this macro generates, to be more specific it is sentence number 5 of 17. This is one of the many sentences this macro generates, to be more specific it is sentence number 6 of 17. This is one of the many sentences this macro generates, to be more specific it is sentence number 7 of 17. This is one of the many sentences this macro generates, to be more specific it is sentence number 8 of 17. This is one of the many sentences this macro generates, to be more specific it is sentence number 9 of 17. This is one of the many sentences this macro generates, to be more specific it is sentence number 10 of 17. This is one of the many sentences this macro generates, to be more specific it is sentence number 11 of 17. This is one of the many sentences this macro generates, to be more specific it is sentence number 12 of 17. This is one of the many sentences this macro generates, to be more specific it is sentence number 13 of 17. This is one of the many sentences this macro generates, to be more specific it is sentence number 14 of 17. This is one of the many sentences this macro generates, to be more specific it is sentence number 15 of 17. This is one of the many sentences this macro generates, to be more specific it is sentence number 16 of 17. This is one of the many sentences this macro generates, to be more specific it is sentence number 17 of 17.

Figure 12.1. `\XParShape`: extended `\parshape` example.

parameters are the following:

- `\parskip`,
- `\parindent`,
- `\everypar`.

Because these values are saved away at the beginning of a paragraph, it does *not* matter for the *current* paragraph what changes are made to these parameters, after TEX started the current paragraph.

2. The *second set of parameters* are used, when TEX *ends a paragraph* and switches back to vertical mode. It is therefore *irrelevant* what values these parameters had at the beginning of a paragraph or how often these values were changed inside the paragraph. Only the values valid at the very end of the paragraph are applied *to the whole paragraph.*

   The following parameters belong to this group:

- `\baselineskip`,
- `\leftskip` and `\rightskip`,
- `\hangindent` and `\hangafter`,
- `\parfillskip`.

The answer of the original question "what happens if `\baselineskip` is changed in the middle of the paragraph" is therefore, whatever value is valid at the end of the paragraph applies to the *whole* paragraph.

To actually see an example relating to what was discussed above you need to look at the source code in Fig. 12.3, p. 114, together with the output produced by this source in Fig. 12.2 on the next page.

12.3 Vertical Material Trickery, `\vadjust`

The `\vadjust` primitive allows the user to put vertical material, while typesetting a paragraph, out into the vertical list, without influencing the typesetting of the current paragraph in any way.

When a `\vadjust {`⟨vertical material⟩`}` instruction is encountered, the specified vertical material is saved by TEX, and the place where the instruction was encountered is marked. Then the paragraph is built as if there were no `\vadjust` instruction. The marker for the `\vadjust` instruction is preserved, however. When the line-breaking algorithm finally writes each line to the current vertical list, the `\vadjust` material is also written to the current vertical list *after* the line was written out containing the `\vadjust` instruction in the first place.

The vertical material of `\vadjust` is inserted between the current and the following line, *before* the interline glue of the two lines is inserted into the main vertical list.

```
\parskip = 12pt plus 2pt minus 1pt
\parindent = 20pt
    {\tt [1]} This is some text, to show some examples of how certain
parameters of \TeX{} can be set to change the shape of
a paragraph. In the first paragraph nothing is changed.
\par
{\parskip = 0pt plus 1pt
 \parindent = 0pt
{\tt [2]} There is some text now, and we changed {\tt\string\parskip}
and the paragraph indentation. Observe that these values were changed
{\it before\/} the text of the paragraph started and so they will be
used.
\par
}

{\leavevmode
 \parskip = 0pt plus 1pt
 \parindent = 0pt
{\tt [3]} There is some text now. We changed {\tt\string\parskip}
and {\tt \string\parindent}. Observe that these values were
changed {\it after\/} \TeX{} started the paragraph
({\tt\string\leavevmode}). We also might just have started the
paragraph and then changed those values. The effect would be the
same.
\par
}

{\leftskip = 3in
 \rightskip = 3in
{\tt [4]} In all the above examples the changes of those parameters
were made local, using curly braces. Here is one final example
where we change {\tt\string\leftskip} and {\tt\string\rightskip}.
The initial values are changed again here
\leftskip = 0.5in \rightskip = 4.0in and finally they are changed
before the paragraph is terminated. \leftskip = 1in
\rightskip = 0.5in Those values are the values finally taken.
\par
}

{\leftskip = 1in
 \rightskip = 1in
{\tt [5]} This example now shows one of the commonly made mistakes.
The values which are valid at the end of a paragraph are the values
which are used for the whole paragraph. The end of the paragraph
is indicated by the end of the line.
But we forgot an empty line before everything was over.
}
\par
{\tt [6]} So the previous empty line was too late!
And so this paragraph is left and right flow to both margins.
```

Figure 12.2. Timing of paragraph parameter changes, source code.

[1] This is some text, to show some examples of how certain parameters of TeX can be set to change the shape of a paragraph. In the first paragraph nothing is changed.
[2] There is some text now, and we changed `\parskip` and the paragraph indentation. Observe that these values were changed *before* the text of the paragraph started and so they will be used.

[3] There is some text now. We changed `\parskip` and `\parindent`. Observe that these values were changed *after* TeX started the paragraph (`\leavevmode`). We also might just have started the paragraph and then changed those values. The effect would be the same.

[4] In all the above examples the changes of those parameters were made local, using curly braces. Here is one final example where we change `\leftskip` and `\rightskip`. The initial values are changed again here and finally they are changed before the paragraph is terminated. Those values are the values finally taken.

[5] This example now shows one of the commonly made mistakes. The values which are valid at the end of a paragraph are the values which are used for the whole paragraph. The end of the paragraph is indicated by the end of the line. But we forgot an empty line before everything was over.

[6] So the previous empty line was too late! And so this paragraph is left and right flow to both margins.

Figure 12.3. Timing of paragraph parameter changes, output.

Let me show you an example. For this example the following input is used:

```
$$
\vbox{
    \hsize = 4in
    In the following paragraph assume a
    \verb+\vadjust{\vskip 6pt}+ is inserted right after
    {\it here\/}\vadjust{\vskip 6pt} (where the word ``here''
    is printed in italics). This means, that the distance
    of the line which contains the \verb+\vadjust+
    and the following line is increased by 6pt.
}
$$
```

The preceding input generates the following output:

In the following paragraph assume a `\vadjust{\vskip 6pt}` is inserted right after *here* (where the word "here" is printed in italics). This means, that the distance of the line which contains the `\vadjust` and the following line is increased by 6pt.

12.3.1 Applications of \vadjust

Let me list some applications of **\vadjust**:

1. Leave space between consecutive lines in a paragraph to insert a figure. This application will be discussed in the next Subsection in more detail.
2. Insert a penalty and/or vertical glue between two lines of a paragraph:
   (a) **\vadjust{\nobreak}** prevents a page break between the current and the next line, where the term "current line" means the line where the **\vadjust** was encountered.
   (b) **\vadjust{\vfill\eject}** does the opposite, forcing a page break in the middle of a paragraph (the line where the instruction occurs is completely filled up with text, because **\vadjust** does not change line breaks in any way). If you do not write **\vfill\eject** inside **\vadjust**, the current text would be split up in two separate paragraphs, because **\vfill** forces TeX into vertical mode.

12.3.2 \vadjust for Inserting Figures

The idea of using **\vadjust{\vskip ...}** can be used to create some white space in the middle of a paragraph to insert a figure. For instance, **\vadjust{\vskip 2.0in}** leaves two inches of white space between the line where this instruction occurs and the next line, and a figure could be glued in between these two lines.

There is a problem though with this approach: if there is *insufficient* space left on the current page (less than 2.0 in), this glue causes a page break. This means that the current page will be spaced out vertically in an ugly manner. Because this glue disappears as part of the page break, neither on the current nor on the next page will 2 in of white space be reserved. In this case (where there is insufficient room), we probably are looking for a different solution. This Subsection discusses a solution where the figure occurs on the top of the next page with text on the current page continuing until the current page is filled up.

The macro I will present now is for the insertion of a figure in the middle of a paragraph. If there is *sufficient* room on the current page, **\vadjust** is used to leave the necessary amount of space on the current page. If there is *insufficient* room, a **\topinsert** is generated (see 35.4.3, p. IV-103, for details on **\topinsert**).

The macro **\FigureInPar** has the following five parameters:

- **#1**. The part of the paragraph, that should *precede* the figure space box reserved by this macro.
- **#2**. The *height* of the *figure space box* to be inserted (rules are drawn to indicate the reserved space). The depth of this box is assumed to be zero.

- #3. How much *vertical space* should be left on the current page, measured from the *bottom* of the figure space box to the *end of the page*. This parameter therefore allows the specification of a minimum number of lines which are printed below the figure space box.
- #4. *Vertical space* inserted *before* and *after the figure space box*, in case the figure is inserted between the lines of a paragraph. This space is allocated to ensure a reasonable spacing between paragraph lines and the figure space box.
- #5. The *second part of the paragraph's text*, the part which is supposed to follow the figure space box.

  Note that #1 and #5 together form the paragraph's text. The part of the paragraph which *precedes* the figure (assuming the figure box becomes a \vadjust and not a \topinsert) is #1, plus a couple of words of the beginning of #5.

The source code for macro \FigureInPar begins here.

$\mathcal{P}'$ • figinpar.tip •

```
\InputD{sumhd.tip}                           % 4.5.10, p. I-102.
\InputD{box-mac.tip}                         % 9.3.14, p. I-343.
\InputD{freespac.tip}                        % 32.2.4, p. IV-6.
\def\FigureInPar #1#2#3#4#5{%
    \par
    \message{\string\FigureInPar: start}%
```

Start a group keeping all changes local.

```
    {%
```

Compute space required by the text of the paragraph preceding the figure box, the figure box itself and any other space which needs to be available.

```
        \setbox 0 = \vbox{#1}
        \OverallSize{\dimen0}{0}%
        \advance\dimen0 by #2
        \advance\dimen0 by #3
```

The space of #4 has to be accounted for twice, because it precedes and follows the figure box, if the figure box will be placed inside the text.

```
        \advance\dimen0 by #4
        \advance\dimen0 by #4
        \advance\dimen0 by \parskip
```

\dimen0 now contains the amount of vertical space which is *required* on the current page. Compute the amount of *available* space and then the difference between these two values. \FreePageSpace is set by \ComputeFreeSpaceOnPage.

If \FreePageSpace is positive, then there is enough room. If the value is negative, the figure box is treated as a \topinsert.

```
        \message{\string\FigureInPar:
            available space: \the\FreePageSpace}%
```

Generate a ruled box to indicate where the figure itself should be glued in later.

```
        \setbox 0 = \HboxR{\EmptyBox{#2}{0pt}{\hsize}}%
```

Now it's time to get down to business: which of the two cases applies?

```
        \ifdim\FreeSpaceConditional < \dimen0
```

There is insufficient space (**\FreePageSpace** is negative), and a **\topinsert** for the figure box is generated. Before that the paragraph itself is printed (the text of which can be found in **#1** and **#5**).

```
            \message{\string\FigureInPar:
                insufficient space: make it a \string\topinsert.}%
```

The following **\unskip** eliminates any trailing space which might be part of **#1**.

```
            #1\unskip
            \space
            #5%
            \par
            \topinsert
                \box0
            \endinsert
        \else
```

There is sufficient space: print the text preceding the figure box, insert the figure box and then print the remaining text of the paragraph.

```
            \message{\string\FigureInPar:
                Sufficient space: put it here.}%
            #1\unskip
            \space
            \vadjust{%
                \vskip #4
                \box0
                \vskip #4
            }%
            #5\par
        \fi
```

End the group started earlier by this macro.

```
    }
}
```

• End of **figinpar.tip** •

And here is an application: the following input will generate output as it is reprinted in Figs. 12.4–12.5, pp. 119–120.

• **ex-figinpar.tip** •

```
\input inputd.tip
\InputD{box-mac.tip}                    % 9.3.14, p. I-343.
\InputD{figinpar.tip}                   % 12.3.2, p. 116.
\InputD{ts-dime3.tip}                   % 31.2.4.3, p. III-598.
```

Print no page numbers.

```
\nopagenumbers
```

The first part of a paragraph.

```
\def\PreText{%
    This is the text, which starts one of our sample paragraphs,
    which will have a figure space inserted in the middle of a
    paragraph.  The end of the text preceding the figure is
    marked with a $\bullet$.
}
```

The second part of a paragraph.

```
\def\PostText{%
    Oh well, that is the text after it, and more and after and even
    more and that is the text after it, and more and after and
    even more and that's all as of now.
    Oh well, that is the text after it, and more and after and even
    more and that is the text after it, and more and after and
    even more and that's all as of now.
    Oh well, that is the text after it, and more and after and even
    more and that is the text.
}
```

Set up some other parameters, then produce some examples.

```
\parskip = 12pt plus 2pt minus 1pt
\vsize = 7.5in

\FigureInPar{\PreText}{2.0in}{0.5in}{5.0pt}{\PostText}
\FigureInPar{\PreText}{1.8in}{0.5in}{10.0pt}{\PostText}
\FigureInPar{\PreText}{2.3in}{0.5in}{5.0pt}{\PostText}
\FigureInPar{\PreText}{1.8in}{0.5in}{5.0pt}{\PostText}
\bye
```

• End of `ex-figinpar.tip` •

12.4 Table of Contents Typesetting

I will now discuss the typesetting of a table of contents. The following discussion also applies to a list of figures and a list of tables. Below I will show my solution to the attempt of defining a very general type of macro which can be adopted to a variety of circumstances.

Because table of contents entries can be longer than one line it is *not* possible to use hboxes to enclose one line of a table of contents entry. Instead each table of contents entry must be typeset as a separate paragraph.

Generating a table of contents file is discussed in 29.2, p. III-499, and not dealt with here. The assumption in this Section is that there is a table of contents file available which contains a series of calls to macro `\TocEntry`. This macro has the following four parameters:

This is the text, which starts one of our sample paragraphs, which will have a figure space inserted in the middle of a paragraph. The end of the text preceding the figure is marked with a •. Oh well, that is the text after it, and more and after and even more and that is the text after it, and more and after and even more and that's all as of now. Oh well, that is the text after it, and more and after and even more and that is the text after it, and more and after and even more and that's all as of now. Oh well, that is the text after it, and more and after and even more and that is the text.

This is the text, which starts one of our sample paragraphs, which will have a figure space inserted in the middle of a paragraph. The end of the text preceding the figure is marked with a •. Oh well, that is the text after it, and more and

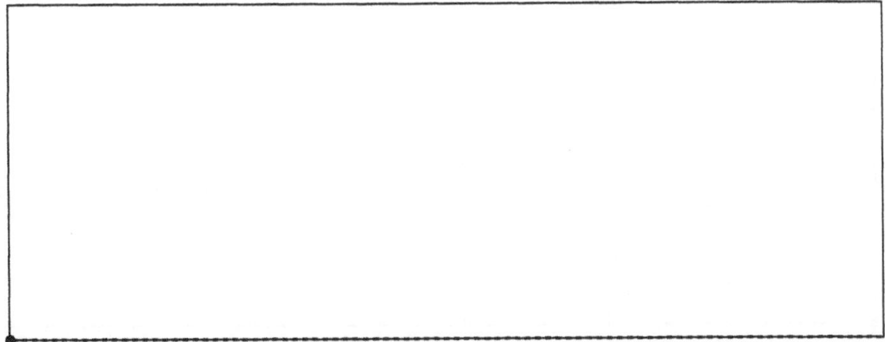

after and even more and that is the text after it, and more and after and even more and that's all as of now. Oh well, that is the text after it, and more and after and even more and that is the text after it, and more and after and even more and that's all as of now. Oh well, that is the text after it, and more and after and even more and that is the text.

This is the text, which starts one of our sample paragraphs, which will have a figure space inserted in the middle of a paragraph. The end of the text preceding

Figure 12.4. First page of example of a figure inserted between the lines of a paragraph.

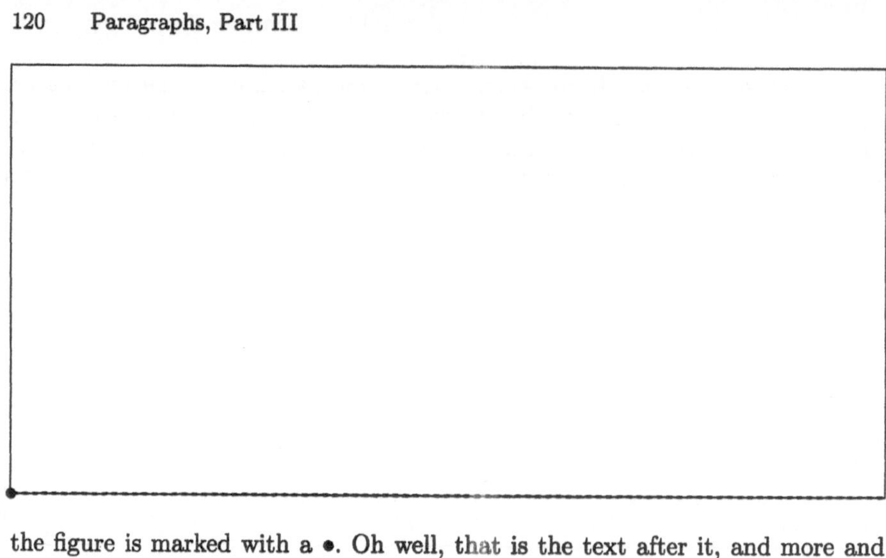

the figure is marked with a •. Oh well, that is the text after it, and more and after and even more and that is the text after it, and more and after and even more and that's all as of now. Oh well, that is the text after it, and more and after and even more and that is the text after it, and more and after and even more and that's all as of now. Oh well, that is the text after it, and more and after and even more and that is the text.

This is the text, which starts one of our sample paragraphs, which will have a figure space inserted in the middle of a paragraph. The end of the text preceding the figure is marked with a •. Oh well, that is the text after it, and more and after and even more and that is the text after it, and more and after and even more and that's all as of now. Oh well, that is the text after it, and more and after and even more and that is the text after it, and more and after and even more and that's all as of now. Oh well, that is the text after it, and more and after and even more and that is the text.

Figure 12.5. Second page of example of a figure inserted between the lines of a paragraph.

- `#1`. The *level* of the table of contents entry. For instance, 0 for the parts of a book, 1 for chapters, 2 for sections, 3 for subsections, etc. This level number has the following two purposes:
  - Determine the layout to be chosen for such a table of contents entry. For instance, one might decide that level 0 and level 1 entries are typeset in boldface with some additional vertical space before and after them, whereas entries with level > 1 are typeset in a regular roman font.
  - Determine whether the entry should be typeset in the first place or not. One, for instance, might decide that in a table of contents only entries down to the section level are listed, and that subsection and subsubsection entries are not printed. The `\TocEntry` macro can suppress such entries under those circumstances (of course, an alternate approach is not to write these entries to the table of contents file in the first place).
- `#2`. The *number* of the heading, that is, for instance, the chapter or section number.
- `#3`. The *text* of the heading.
- `#4`. The *page number* of the heading.

The macro `\TocEntry` must be provided by the user (see below for an example). Next we will discuss the macro `\GenTocEntry`. The macro `\TocEntry` does nothing else than call `\GenTocEntry` to actually achieve the typesetting of table of contents entries.

12.4.1 The Definition of Macro `\GenTocEntry`

The name `\GenTocEntry` is an abbreviation for generic table of contents entry printing macro. This macro has 12 parameters (actually TEX does not allow for that many parameters, and therefore the technique described in 22.3.2, p. III-219, is used).

Any distances "from the left margin" specified below describe distances from the left margin of the standard text of your document, that is text printed with a `\leftskip` value of zero, which is usually the case. Similarly, there are distances specified "from the right margin." The term *continuation lines* means the second, third and so forth line of a "table of contents entry paragraph," thus assuming the first line of such a listing wraps around. Note that, as far as TEX is concerned, every table of contents entry is printed as a separate paragraph, even if this paragraph is only one line long.

- `#1`. The level of the entry.
- `#2`. Distance of the heading number from the left margin of the text. Heading numbers are printed left flush. For the outermost level of the printing of table of contents entries this value is usually zero. It can be increased for entries with higher level numbers, so that these entries are indented on the left hand side.

- #3. Distance of the beginning of the heading's text itself from the left margin of the text. Therefore the heading's number must *not* be longer than the difference of #3 and #2, which is the available space for printing the heading's number.
- #4. Distance of the beginning of continuation lines from the left margin of text. If the same as #3, then continuation lines are flush with all the first lines of table of contents listings.
- #5. Minimum distance between the end of a line of a table of contents entry paragraph, if this line does *not* contain the heading's page number and the right text margin.
- #6. Distance between the right end of a page number in a table of contents listing and the right text margins.
- #7. The distance between the "leader dots" that are the leaders inserted between the table of contents entry and the page number. Specify a zero dimension here, if you do not want any leaders.
- #8. The minimum length of leader dots. If no leader dots are requested (#7 is zero), then this parameter is ignored.
- #9. The number of the heading of the table of contents entry being printed.
- #10. The text of a table of contents entry. See also the next parameter.
- #11. The font to be used to print the table of contents entry. The font specified as this parameter is used for the heading number, the table of contents entry itself, the leader dots and the page number of the table of contents entry. In case you want to use a special font for the table of contents entry text only, make this font specification part of #8. This parameter can be empty in case you want to use the default font of your document.
- #12. The page number of the table of contents entry.

The definition of `\GenTocEntry` has *four* parameters which leaves another eight parameters to be taken care of which are #1 through #8 of `\@GenTocEntry`, a macro defined below and invoked at the very end of `\GenTocEntry`.

$\mathcal{P}'$ • toc-mac.tip •

```
\catcode`\@ = 11
```

Save the level of the current entry in the following counter register.

```
\newcount\@GenTocEntryLevel
```

Here the macro definition starts.

```
\def\GenTocEntry #1#2#3#4{%
      \par
```

Start a group to keep all changes local.

```
      \bgroup
```

Save the level of the current entry.

```
      \global\@GenTocEntryLevel = #1
```

Now set up the various paragraph parameters.

```
    \leftskip = #4
```

Compute the paragraph indentation. The heading's number needs to be printed at the beginning of the line so the paragraph's indentation needs to be set up properly for that.

```
    \parindent = #2
    \advance\parindent by -#4
```

Now set `\dimen0` to the length of the box containing the heading's number.

```
    \dimen0 = #3
    \advance\dimen0 by -#2
```

Continue expansion with macro `\@GenTocEntry`, which will read in another eight parameters.

```
    \@GenTocEntry
}
```

The macro `\@GenTocEntry` continues the work of `\GenTocEntry`. What was described as parameter #i of `\GenTocEntry` is parameter #$i - 4$ of `\@GenTocEntry`.

```
\def\@GenTocEntry #1#2#3#4#5#6#7#8{%
```

Continue setting up other paragraph layout parameters. `\rightskip` is set to a value with generous stretchability so that TeX sets table of contents entries ragged right. Usually table of contents entries are one line long. Two lines occur sometimes, but three and more lines are highly unusual.

```
    \rightskip = #1 plus 1in
```

Next `\parfillskip` is computed as the difference between the right margins for continuation lines and those lines with page numbers in them. Usually **#6** of `\GenTocEntry` is zero, then `\parfillskip` is set to −**#5** of `\GenTocEntry`; this makes the page number line right flush with the standard text margin.

```
    \parfillskip = #2
    \advance\parfillskip by -#1
```

Change to the appropriate font.

```
    #7
```

Build a box which contains one leader character, a centered period. If no leaders are requested (**#7** of `\GenTocEntry` is zero), this box is cleared out.

```
    \ifdim #3 > 0pt
        \setbox0 = \hbox to #3{\hfil.\hfil}
    \else
        \setbox0 = \box\voidb@x
    \fi
```

Now actually typeset the table of contents entry. Start the paragraph first.

```
    \leavevmode
```

Now print the header's number.

```
        \hbox to \dimen0 {#5\hfil}%
```

Print the table of contents entry itself followed by leaders. The `\unskip` specification removes any trailing space from the heading's text (if there is such). The `\nobreak` prevents the following leader from disappearing in a line break. `\leaders\copy0` prints the leader dots. Finally note the glue specification that determines the leader to be printed: the minimum length of leader dots printed is 0.5 in, and it can be as long as necessary.

```
        {%
            #6%
            \unskip
        }%
        \ifvoid 0
            \hfill
        \else
            \nobreak\leaders\copy0\hskip #4plus 1fil
        \fi
```

Next print the page number.

```
        #8%
```

End the table of contents entry paragraph.

```
        \par
```

End the group in which printing this paragraph was enclosed. This group was started in `\GenTocEntry`.

```
        \egroup
    }
    \catcode`\@ = 12
```

• End of `toc-mac.tip` •

12.4.2 An Example

The following example shows how you would usually interface printing a table of contents to reading in a table of contents file. We assume that the table of contents file contains calls to macro `\TocEntry`. This macro `\TocEntry` has the following four parameters (this macro does nothing else than to prepare a call to macro `\GenTocEntry`). You must adjust the definition below to fit your own needs. The macros' four parameters are discussed on p. 118.

• `tocm2.tip` •

```
\InputD{toc-mac.tip}                        % 12.4.1, p. 122.
\def\TocEntry #1#2#3#4{%
    \par
    \ifcase #1
        \vskip 10pt plus 2pt minus 1pt
        \GenTocEntry{0}{0pt}{30pt}{20pt}{1in}{0.5in}{0pt}%
```

```
            {}{#2}{#3}{}{#4}%
    \or
        \GenTocEntry{1}{10pt}{40pt}{40pt}{1in}{0.5in}{5pt}%
            {0.5in}{#2}{#3}{}{#4}%
    \or
        \GenTocEntry{2}{20pt}{50pt}{50pt}{1in}{0.5in}{5pt}%
            {0.5in}{#2}{#3}{}{#4}%
    \else
        \errmessage{\string\TocEntry: nesting too deep.}%
    \fi
}
```

• End of `tocm2.tip` •

Here is the example source code.

• `ex-toc.tip` •

```
\input inputd.tip
\InputD{tocm2.tip}                              % 12.4.2, p. 124.
\TocEntry{0}{1}{This is fun}{1}
\TocEntry{1}{1.1}{This is a lot of fun}{2}
\TocEntry{1}{1.2}{This is a lot of fun and this is even more}{3}
\TocEntry{1}{1.3}{This is a lot of fun and this is even more than}{4}
\TocEntry{1}{1.4}{This is a lot of fun and this is even more
    fun than before}{5}
\TocEntry{1}{1.5}{This a lot, a lot of fun and this is even
    fun than before and}{6}
\TocEntry{1}{1.6}{This is a lot, a lot of fun and this is even
    fun than before and more}{7}
\TocEntry{1}{1.7}{This is a lot, a lot of fun and this is even
    fun than before and more like}{8}
\TocEntry{1}{1.8}{This is a lot, a lot of fun and this is even
    fun than before and more like this}{9}
\TocEntry{1}{1.9}{This is a lot, a lot of fun and this is even
    fun than before and more like this and}{10}
\TocEntry{1}{1.10}{This is a lot, a lot of fun and this is even
    fun than before and more like this and like this}{11}
\TocEntry{1}{1.11}{This is a lot, a lot of fun and this is even
    fun than before and more like this and like this more}{12}
\TocEntry{1}{1.12}{This is a lot, a lot of fun and this is even
    fun than before and more like this and like this more likes it
    it it it it}{18}
\TocEntry{1}{1.18}{This is a lot of fun}{19}
\TocEntry{1}{1.2}{This is a lot of fun and when we have a lot
    of fun then we should have more than a lot of fun}{20}
\TocEntry{1}{1.3}{This is a lot of fun and when we have a lot
    of fun then we should have more than a lot of fun and so on}{21}
\TocEntry{1}{1.4}{This is a lot of fun and when we have a lot
    of fun then we should have more than a lot of fun and and}{22}
\TocEntry{1}{1.5}{This is a lot of fun and when we have a lot
    of fun then we should have more than a lot of fun
    and as we go one we may find more interesting things}{23}
\TocEntry{1}{1.6}{This is a lot of fun and when we have a lot
```

```
    of fun then we should have more than a lot of fun.
    This is an unbearable title}{27}
\TocEntry{2}{1.6.1}{This is a lot of fun and when we have a lot
    of fun and more then we should have more than a lot of fun
    of fun and more then we should have more than a lot of fun}{100}
\TocEntry{2}{1.6.2}{This is a lot of fun and when we have a lot
    of fun and more then we should have more than a lot of fun
    of fun and more then we should have more than a lot of fun
    of fun and more then we should have more than a lot of fun}{103}
\TocEntry{2}{1.6.3}{This is a lot of fun and when we have a lot
    of fun and more then we should have more than a lot of fun
    of fun and more then we should have more than a lot of fun
    of fun and more then we should have more than a lot of fun
    of fun and more then we should have more than a lot of fun}{105}
\TocEntry{2}{1.6.4}{This is a lot of fun and when we have a lot
    of fun and more then we should have more than a lot of fun}{200}
\TocEntry{0}{2}{This is a lot of fun and when we have a lot
    of fun then and more and so one and
    more as you like we should have more than a lot of fun}{256}
\TocEntry{0}{3}{This is a lot of fun and when we have a lot
    of fun then we should this is and so
    well have you may be and more and more and
    have more than a lot of fun}{300}
\TocEntry{0}{4}{This is a lot of fun and when we have a lot
    This is incredible because when we like it very much then
    it may be actually too long, of fun then we should have
    more than a lot of fun}{423}
\bye
```

• End of ex-toc.tip •

The above source generates the following output:

1 This is fun 1

1.1 This is a lot of fun 2

1.2 This is a lot of fun and this is even more 3

1.3 This is a lot of fun and this is even more than 4

1.4 This is a lot of fun and this is even more fun than before 5

1.5 This a lot, a lot of fun and this is even fun than before and 6

1.6 This is a lot, a lot of fun and this is even fun than before and more 7

1.7 This is a lot, a lot of fun and this is even fun than before and more like 8

1.8 This is a lot, a lot of fun and this is even fun than before and more like this 9

1.9 This is a lot, a lot of fun and this is even fun than before and more like this and 10

1.10 This is a lot, a lot of fun and this is even fun than before and more like this and like this 11

be actually too long, of fun then we should have more than a lot of fun 423

12.4.3 Table of Contents Printing for This Series

The table of contents entry macros of 30.7.5, p. III-555, generate a table of contents file with calls to a macro **\EntryIntotoc**. This macro is defined next. It is used to print the table of contents of the four volumes of this series.

$\mathcal{P}'$ • **ts-toc.tip** •

```
\InputD{toc-mac.tip}                             % 12.4.1, p. 122.
```

This macro has the following parameters:

- **#1**. The level of the nesting (chapter level is 1, section level is 2, subsection level is 3 and subsubsection level is 4).

  Level 0 is used for subdivisions within the endmatter which are unnumbered (for instance, the "General Comments on TEX Practice," the index, the listing of trademarks, the bibliography, and the acknowledgements of this series).
- **#2**. The number of the heading.
- **#3**. The heading's text.
- **#4**. The page number.

```
\def\EntryIntotoc #1#2#3#4{%
    \par
```

Any additional vertical space to insert? On the chapter level (**#1** = 1) throw in an extra blank line. Also between entries of the end matter (**#1** = 0).

```
    \ifcase #1
        \bigskip
    \or
        \bigskip
    \else
```

No extra space in other cases.

```
        \relax
    \fi
```

Now print the table of contents entry itself.

```
    \ifcase #1
```

End matters, like bibliography, index, etc. In this case **#2** is empty.

```
        \GenTocEntry{0}{0pt}{20pt}{20pt}{30pt}%
            {0pt}{5pt}{0.5in}{#2}{#3}{\rm}{#4}
    \or
```

Chapter level (level 1).

```
            \GenTocEntry{1}{0pt}{40pt}{40pt}{30pt}%
                {0pt}{10pt}{0.5in}{#2}{#3}{\bf}{#4}
        \or
```

Section level (level 2).

```
            \GenTocEntry{2}{10pt}{50pt}{50pt}{30pt}%
                {0pt}{5pt}{0.5in}{#2}{#3}{\rm}{#4}
        \or
```

Subsection level (level 3).

```
            \GenTocEntry{#1}{20pt}{60pt}{60pt}{30pt}%
                {0pt}{5pt}{0.5in}{#2}{#3}{\rm}{#4}
        \or
```

Subsubsection level (level 4). All those entries are simply dropped.

```
        \else
```

No level beyond the subsubsection level is used in this series.

```
            \errmessage{\string\EntryIntotoc: no level #1 subdivision
                in this series.}
        \fi
    }
```

• End of `ts-toc.tip` •

12.4.4 List of Figure and List of Table Printing for This Series

The main difference between the definitions for the list of figure and list of table printing for this series compared to the preceding Subsection's macro definition is that there are no different *levels* of entries. All entries in a list of figures or tables are of level 1. On the other hand, we want to separate vertically entries of *different* chapters. For that purpose level 0 entries are used (these are no real entries; the macro below simply uses a level 0 entry to generate that additional vertical space).

Here is the definition of `\EntryIntolof`. The macro `\EntryIntolot` for the list of table entries is identical (both macros have the same parameters as `\EntryIntotoc` of the preceding Subsection).

$\mathcal{P}'$ • `ts-loft.tip` •

```
\InputD{toc-mac.tip}                        % 12.4.1, p. 122.
\InputD{vsmax.tip}                          % 8.6, p. I-310.
\def\EntryIntolof #1#2#3#4{%
```

Additional vertical space is inserted between listings of different chapters.

```
        \ifcase #1
```

Using `\MaxVskip` multiple `\MaxVskip` calls do not lead to vertical glue being accumulated. This is necessary so it is possible to have a sequence of chapters without tables or figures in them.

```
        \MaxVskip{12pt}%
    \or
        \GenTocEntry{#1}{10pt}{50pt}{50pt}{25pt}%
            {0pt}{5pt}{0.5in}{#2}{#3}{\rm}{#4}
    \else
        \errmessage{\string\EntryIntolof/lot: illegal level.}%
    \fi
}
```

Now define `\EntryIntolof`.

```
\let\EntryIntolot = \EntryIntolof
```

• End of `ts-loft.tip` •

12.5 `\prevgraf`

`\prevgraf` is a counter parameter which at the beginning of a paragraph is set to zero (see 10.4.2, p. 12). At the end of a paragraph it is set to the number of lines in that paragraph. This parameter is also set and changed when a paragraph is temporarily suspended and a formula in display math mode is started. We are not interested in this aspect right now.

12.5.1 A Macro To Determine the Number of Lines of a Paragraph, `\GetNumberOfLines`

Here we show briefly a macro `\GetNumberOfLines` which determines the number of lines in a paragraph and which uses `\prevgraf`.

$\mathcal{P}'$ • `getnuml.tip` •

The macro `\GetNumberOfLines`, which is defined below, stores its result in counter register `\ResultNumberOfLines`, which is declared next.

```
\newcount\ResultNumberOfLines
```

The macro `\GetNumberOfLines` determines the number of lines of a paragraph. After calling this macro the result can be picked up in `\ResultNumberOfLines`. This macro has the following parameters:

- `#1`. The text of the paragraph.
- `#2`. The width at which it is supposed to be set.

```
\def\GetNumberOfLines #1#2{%
```

Start a group.

```
        {%
```

Store the paragraph in a vbox.

```
            \setbox 0 = \vbox{%
```

Set **\hsize** to the desired value.

```
                \hsize = #2
                #1
```

End the paragraph.

```
                \par
```

Save **\pregraf**.

```
                \global\ResultNumberOfLines = \prevgraf
            }%
        }%
}
```

• End of **getnuml.tip** •

12.5.2 Setting \prevgraf

The user actually has the possibility to set **\prevgraf** to any value—line break decisions are not affected unless paragraph shapes with **\parshape** or **\hangindent** are generated. This is what we are going to look into now.

If you read 10.4.2, p. 12, carefully then you discover that **\prevgraf** is reset at the beginning of a paragraph. The following input

• **ex-xparshape-p.tip** •

```
\input inputd.tip
\InputD{samplepa.tip}                          % 27.1.3.3, p. III-402.
\InputD{x-parsh.tip}                           % 12.1.3, p. 108.
\XParShape = 3
    10  30pt    0pt     205pt   5pt
    10  30pt    5pt     255pt   -5pt
    10  80pt    -5pt    200pt   10pt
    \noindent
    \prevgraf = 7
    \SamplePar{X}{17}
```

• End of **ex-xparshape-p.tip** •

generates Fig. 12.6 on the next page. The **\noindent** instruction was necessary to start the paragraph, but before any text is contributed **\prevgraf** is set to 7. Thus the first line really becomes the seventh line. Compare the output of this figure with the output of Fig. 12.1, p. 111.

Identification of this paragraph: *X. Sample paragraph 2, with 17 sentences.* So here we go, and when you check the number of sentences, then note that these first two sentences do *not* count. This is one of the many sentences this macro generates, to be more specific it is sentence number 1 of 17. This is one of the many sentences this macro generates, to be more specific it is sentence number 2 of 17. This is one of the many sentences this macro generates, to be more specific it is sentence number 3 of 17. This is one of the many sentences this macro generates, to be more specific it is sentence number 4 of 17. This is one of the many sentences this macro generates, to be more specific it is sentence number 5 of 17. This is one of the many sentences this macro generates, to be more specific it is sentence number 6 of 17. This is one of the many sentences this macro generates, to be more specific it is sentence number 7 of 17. This is one of the many sentences this macro generates, to be more specific it is sentence number 8 of 17. This is one of the many sentences this macro generates, to be more specific it is sentence number 9 of 17. This is one of the many sentences this macro generates, to be more specific it is sentence number 10 of 17. This is one of the many sentences this macro generates, to be more specific it is sentence number 11 of 17. This is one of the many sentences this macro generates, to be more specific it is sentence number 12 of 17. This is one of the many sentences this macro generates, to be more specific it is sentence number 13 of 17. This is one of the many sentences this macro generates, to be more specific it is sentence number 14 of 17. This is one of the many sentences this macro generates, to be more specific it is sentence number 15 of 17. This is one of the many sentences this macro generates, to be more specific it is sentence number 16 of 17. This is one of the many sentences this macro generates, to be more specific it is sentence number 17 of 17.

Figure 12.6. `\XParShape`: modified extended `\parshape` example.

12.6 Now It Is Playtime

I will now present some more examples which show various combinations of setting TEX's parameters to typeset paragraphs. These different examples should give you a better feeling for the various paragraph-related parameters.

12.6.1 Hanging Indentation, Margins Moved In

The following example shows that **\leftskip** and **\rightskip** can be used to move in the margins of a paragraph simultaneously with hanging indentation generated by assigning proper values to **\hangindent** and **\hangafter**.

The input

```
\leftskip = 20pt
\rightskip = 30pt
\hangindent = 40pt
\hangafter = 2
\noindent
This example shows that by changing {\tt\string\leftskip} and
{\tt\string\rightskip} and assigning proper values to
{\tt\string\hangindent} and {\tt\string\hangafter} it is
possible to have hanging indentation with both margins moved in.
This example is another example which demonstrates the enormous
flexibility of \TeX.
```

generates the following output:

This example shows that by changing **\leftskip** and **\rightskip** and assigning proper values to **\hangindent** and **\hangafter** it is possible to have hanging indentation with both margins moved in. This example is another example which demonstrates the enormous flexibility of TEX.

12.6.2 Hanging Indentation, Ragged Right

Let me show hanging indentation with a ragged right type of paragraph layout. This can be achieved very easily. We simply combine the discussion of hanging indentation (11.4, p. 80) with the discussion of ragged right text (11.6.1, p. 99).

The following source code was used:

```
\leftskip = 0.5in
\rightskip = 0.5in plus 6em
\spaceskip = 0.33333em
\xspaceskip = 0.5em
\hangindent = 40pt
\hangafter = 2
    This paragraph uses a ragged right type of text, with hanging
indentation. There is not too much of a difference compared to the
paragraph in the previous example. We need some more text here,
so the paragraph becomes longer and you get indeed a clear impression
of what is going on.
```

This generates the following output:

This paragraph uses a ragged right type of text, with hanging indentation. There is not too much of a difference compared to the paragraph in the previous example.

We need some more text here, so the paragraph becomes longer and you get indeed a clear impression of what is going on.

12.6.3 Showing Bad Typesetting

In the following example let me emulate a cheap word processor. For that purpose we turn off hyphenation, use long words, and give the interword glue lots of stretchability. The margins are moved in by using `\leftskip` and `\rightskip` to generate a rather narrow column. Also `\parfillskip` is set to zero so that the last line will be filled with text all the way (and not appear ragged right).

The following input was used:

```
\leftskip = 1.5in
\rightskip = 1.5in
\spaceskip = 0.33333em plus 6em
\xspaceskip = 0.5em plus 8em
\parfillskip = 0pt
\hyphenpenalty = 10000
\tt
Here we simulate a poor word processor. Mathematics, Electrical
Engineering and other words like probability, stretchability and
shrinkability are very long words, and you easily see
that \TeX{} is even able to emulate a poor word processor.
```

This input generates the following output:

```
     Here   we   simulate   a
poor    word    processor.
Mathematics,    Electrical
Engineering    and    other
words  like  probability,
stretchability         and
shrinkability   are   very
long   words,   and    you
easily  see  that  TeX  is
even   able   to   emulate
a  poor  word  processor.
```

12.6.4 First Line Special, `\FirstLineSpecial`

Let me now discuss a macro `\FirstLineSpecial` which typesets the first line of a paragraph using a font different from the font used for the rest of the paragraph. This macro has one parameter, `#1`, which is the font instruction for the first line of the paragraph. The text of the paragraph itself should follow the call of `\FirstLineSpecial`, without an empty line between the argument of a `\FirstLineSpecial` call and the paragraph's text.

$\mathcal{P}'$ • **firstldi.tip** •

```
\catcode`\@ = 11
\def\FirstLineSpecial #1{%
```

If there is any preceding paragraph not yet typeset, then do so now.

```
    \par
```

Start a group and switch to the requested font for the first line.

```
    \begingroup
    #1

    \def\@FLDTemp{}%
```

Store in `\dimen0` the value of `\parindent` minus the width of one space. The reason for subtracting the width of one space is that word after word is collected each time with a trailing space after each word (see the definition of `\@FLDTempa` inside `\@FLDThree`), but with n words on a line, there are only $n-1$ interword spaces on that line.

```
    \dimen0 = \parindent
    \setbox0 = \hbox{ }%
    \advance\dimen0 by -\wd0
    \@FLDOne
}
```

The following macro picks up the next word (delimited by a space) and inserts it plus a space into macro `\@FLDTempa`. This macro is called to pick up either the very first word or to pick up subsequent words, if the line so far was determined too long.

```
\def\@FLDOne #1 {%
    \wlog{\string\@FLDOne: called with "#1".}%
    \xdef\@FLDTempa{#1\ }%
```

Start determining the length of the material collected so far.

```
    \@FLDTwo
}
```

The following macro is called after yet another word was absorbed.

```
\def\@FLDTwo{%
    \wlog{\string\@FLDTwo: called}%
```

Measure the length of the word just collected, add it to the length of the text so far collected.

```
    \setbox0 = \hbox{\@FLDTempa}%
    \advance\dimen0 by \wd0
```

If the line is still too short, add this word to the words collected so far and collect the next word, that is restart the correction.

```
    \ifdim\dimen0 < \hsize
```

```
            \edef\@FLDTemp{\@FLDTemp\@FLDTempa}%
            \let\@FLDNext = \@FLDOne
        \else
```

What you have collected is enough. Start the paragraph and print the text collected so far.

```
            \leavevmode
            \@FLDTemp
```

Remove the trailing space after the last word on the first line of the paragraph's text and then force a line break.

```
            \unskip
            \break
```

After termination of the group that enclosed the font change is terminated, the most recent word (which is the first word of the second line of the paragraph) is printed.

```
            \aftergroup\@FLDTempa
            \let\@FLDNext = \endgroup
        \fi
```

Here either the recursion is restarted to collect the next word, or the rest of the paragraph's text is being typeset.

```
        \@FLDNext
}
\catcode'\@ = 12
```

• End of firstldi.tip •

Here is an example of the application of this macro. The following input was used:

• ex-firstldi.tip •

```
\input inputd.tip
\InputD{firstldi.tip}                        % 12.6.4, p. 135.
```

Load a 10 pt roman all caps font for the first line of the following paragraph.

```
\font\capsfont = cmcsc10
$$
```

Typeset the paragraph here.

```
    \vbox{
        \hsize = 4in
        \FirstLineSpecial{\capsfont}%
        This is a wonderful example of how this macro is
        supposed to work, so good luck is all I can wish you.
        Let's make the sample text a little longer and
        then let's leave it at that.
    }
$$
```

• End of ex-firstldi.tip •

The preceding input generates the following output:

> THIS IS A WONDERFUL EXAMPLE OF HOW THIS MACRO IS supposed to work, so good luck is all I can wish you. Let's make the sample text a little longer and then let's leave it at that.

The preceding macro is by no means restricted to the typesetting of the first line of a paragraph in small caps. You can use any other font for the first line of a paragraph.

12.7 The Line Breaking Algorithm and Hyphenation

In this Section we will go into much more detail and analyze the functioning of TEX's way to compute the line breaks. We will summarize all the parameters which control line breaking and show various examples. The following discussion is important to understand what can be done to typeset narrow columns the subject of another Section.

Some examples will show how the positioning of line breaks influences the appearance of a paragraph. This is to allow the user when it is occasionally necessary to adjust these parameters.

12.7.1 Tracing Line Break Computations, `\tracingparagraphs`

In TEX there is the possibility of tracing its line break computations by setting `\tracingparagraphs` to a positive value (the default is zero in which case no computations are shown). In this series we will not go into that much detail as far as line break computations are concerned; see the TEXbook for details.

12.7.2 A Two Pass Algorithm

Computing the line breaks is performed in two passes by TEX.

1. *Pass 1.* Line breaks are attempted *without* hyphenation. This pass will fail, if the badness of any line *exceeds* the value of `\pretolerance`, and in this case pass 2 will be attempted. See 5.3, p. I-127, for an explanation of the term badness.

   Let me discuss possible values of `\pretolerance`:

   (a) The default of `\pretolerance` in the plain format is 100.

   (b) If you set `\pretolerance = 10000`, then the first pass will succeed in almost all cases (spacing may not be very good though), because you told TeX that any badness is acceptable. Pass 2 will never be started and no lines will be hyphenated.
   (c) If you set `\pretolerance = -1` then the opposite happens: pass 1 will not even be started, and TeX will start with pass 2 right away.

2. *Pass 2.* Try line breaks *with* hyphenation. The value of `\tolerance` determines the maximum badness lines can have in this case. The default value of this parameter in the plain format is 100.

The reason for subdividing the line breaking into two passes is that pass 1, which will succeed in many cases, is much faster to execute. So it is worth the risk of trying pass 1 first, even if one has to do pass 2 later because pass 1 did not succeed. Look through this volume, and you can see for yourself, that hyphenation does not happen very frequently (especially in a book like this one, where many paragraphs are fairly short).

12.7.3 The `\showhyphens` Command

Before discussing hyphenation, let me present macro `\showhyphens` from the plain format. This macro displays the hyphenation of a word in TeX's log file. Therefore if you wanted to find out how TeX would hyphenate a word, you will not need to write some artificial text to see how TeX hyphenates a specific word. Here is an example source file:

• `ex-showhyphens.tip` •

```
\showhyphens{elementary}
\showhyphens{paragraphs}
\showhyphens{complication}
\bye
```

• End of `ex-showhyphens.tip` •

This source code generates the following log file:

• `ex-showhyphens.log` •

```
This is TeX, C Version 3.14 (...)
**&/usr/local/tex/lib/fmt/plain ex-showhyphens.tip
(ex-showhyphens.tip
Underfull \hbox (badness 10000) in paragraph at lines 1--1
[] \tenrm el-e-men-tary

\hbox(6.94444+1.94444)x16383.99998, glue set 9787.77635 []

Underfull \hbox (badness 10000) in paragraph at lines 2--2
[] \tenrm para-graphs

```

```
\hbox(6.94444+1.94444)x16383.99998, glue set 9787.39299 []

Underfull \hbox (badness 10000) in paragraph at lines 3--3
[] \tenrm com-pli-ca-tion

\hbox(6.94444+1.94444)x16383.99998, glue set 9783.12628 []

)
No pages of output.
```

12.7.4 How TeX Determines the Hyphenation of a Word, `\-`, `\discretionary`

The following discussion of how TeX determines the hyphenation of a specific word will include general details of the line breaking algorithm.

1. TeX hyphenates only in pass 2 of the line breaking algorithm. The following discussion is therefore mainly of interest if pass 2 of the line breaking algorithm is actually executed.
2. If a word (or for that matter words) is enclosed inside an hbox, then it cannot be hyphenated by TeX, because an hbox is regarded as one solid unit which cannot be broken up.
3. If a word contains one or more *discretionary hyphens* \- then TeX uses those discretionary hyphens *only*. TeX does not use any of the built-in rules nor does it combine these built-in rules (discussed below) with the discretionary hyphens.

   TeX will consider hyphenation at the places marked by discretionary hyphens. The main purpose of discretionary hyphens is to give the user an opportunity to overwrite TeX's built-in rules. For instance, if you entered **abc\-def**, TeX would, if it found hyphenation appropriate, hyphenate the word **abcdef** right in the middle, and nowhere else.

   Note that if you corrected the hyphenation of a word using discretionary hyphens, then TeX will *not* remember the discretionary hyphens if it encounters the same word later, so you need to repeat discretionary hyphens.
4. By the way, \- is a primitive. If this primitive were not provided, you could define a macro \- quite easily as follows:

```
\def\-{%
    \discretionary{-}{}{}%
}
```

   We therefore need to discuss `\discretionary` next. This primitive is used to indicate discretionary hyphens, but it is more general than \-. `\discretionary` has three arguments (it's nevertheless a primitive):

   - #1. The *pre-break* text.

- #2. The *post-break* text.
- #3. The *nobreak* text.

The text `xx\discretionary{a-}{b}{cc}yy` is interpreted by TeX as follows:

(a) There is the possibility ("discretionary hyphen") of hyphenation after `xx`.
(b) If TeX hyphenates at this point, then `xx` is printed followed by `a-`, the pre-break text, on the current line; the following line starts with `b`, the post-break text, followed by `yy`.
(c) If TeX does not need to hyphenate, it prints the nobreak text, that is `xxccyy` is printed.

The above instruction obviously gives you much more control over hyphenation then you usually need, but there are cases in many languages other than English where such a fine control is needed. For instance, the German word `Bettuch`, when hyphenated, suddenly has three `t`s in it, as in `Bett-tuch`. To get the correct hyphenation for this word in TeX you would have to write:

```
Be\discretionary{tt-}{t}{tt}uch
```

Here is one special case: if you write `\discretionary{}{}{}` in the middle of a word, then this means that you can break this word right at this place, but the break will not be visible in any way (with the exception of the fact that the parts of the word preceding and following the `\discretionary{}{}{}` construct appear on separate lines).

If discretionary hyphens are inserted, TeX does not apply its regular hyphenation rules.

5. TeX is *multilingual.* The current language is determined by an integer value, stored in the counter parameter `\language`. Here are the rules of how the integer stored in this register is to be interpreted:

(a) You, the user, are responsible for assigning a unique integer for every language you use.
(b) Values in the range of 0 to 255 are permissible, in other words hyphenation rules of up to 256 different languages are possible in one document.
(c) If `\language` is negative or greater than 255, then TeX acts as if its value were zero.
(d) The default of this register is zero.

Here is a short example. Assume language 0 is English and language 1 is German, here is how you would have to enter some text in both languages.

```
\language = 0
The English sentence ``I have lots of fun to drive an engine''
translates to German as follows:
``{\language = 1 Es macht mir grossen Spass,
        mit einer Lokomotive zu fahren.}''
```

6. To simplify the administration of different language the plain format defines a macro **\newlanguage**. Naturally you need to synchronize the loading of hyphenation patterns with the proper setting of **\language**.

   If you enter, for instance, **\newlanguage\German**, then to switch to the hyphenation patterns of the German language, you need to write **\language = \German**. Writing **\German** by itself is insufficient.

7. The user can change the hyphenation for individual words on a global basis. This way entering discretionary hyphens for the same word is avoided. This is done with **\hyphenation** as in **\hyphenation{abc-def}**. **\hyphentation** is a primitive.

   TeX remembers the hyphenation of this particular word throughout the rest of the document, *but* only in the language which was active when the **\hyphenation** command was issued. Thus **\hyphenation** is also language sensitive.

   For instance, if I wanted to instruct TeX to hyphen the word "**abcdef**" correctly in both languages, English and German, I would have to enter the following instructions:

```
\EnglishLanguage     \hyphenation{abc-def}
\GermanLanguage      \hyphenation{abc-def}
```

   To instruct TeX to not hyphenate a word, simply use **\hyphenation** with no hyphens as in **\hyphenation{word}**.

8. Hyphenation requires some minimum numbers of character preceding and following where hyphenation occurred.

   (a) The parameter register **\lefthyphenmin** (plain format default is 2) determines the minimum number of characters that the fragment of a word must have, *preceding* the hyphenation. For instance, the word **abcdef** could only be hyphenated after the **a**, if the value of this register is either 0 or 1.

       This parameter is a new feature of TeX 3.0.

   (b) The parameter **\righthyphenmin** (plain format default is 3) determines the minimum number of characters for a word fragment must have that *follows* the hyphenation. For instance, the word **abcdef** could only be hyphenated after the **d**, if the value of this register is 2 or less.

       Needless to say this parameter is also a new feature of TeX 3.0.

   (c) If the sum of these two registers is 63 or more, no hyphenation takes place.

9. Finally, if TeX was not able to locate any hyphenation exception for a word, TeX consults its built-in rules.

   (a) These rules are obviously language sensitive and in multilingual documents you must make sure that **\language**'s value is correct.

   (b) The hyphenation rules *cannot* be changed in the middle of a document and they even cannot be loaded in the very beginning, before the processing of a document begins. You must run **initex** to load hyphenation patterns. Because hyphenation patters are language sensitive you need to know which languages you might use in a document.

virtex, which is used to process documents with TEX, does not have the capability of loading hyphenation patterns.

12.7.5 Making the Plain Format Multilingual

If you look into the source code of the plain format (usually stored in a file called `plain.tex`), then you will find there towards the end of this file the following code:

```
% Hyphenation, miscellaneous macros, and initial values
% for standard layout.
\message{hyphenation}

\input hyphen
```

The file `hyphen.tex` is supposed to contain the hyphenation patterns of your "main language."

I changed this part of `plain.tex`. Assume that `hyphen.tex` contains the hyphenation patterns of the English language, and `germanhyphen.tex` the hyphenation patterns on German:

```
\def\English{\language = 0\relax}
\English
\input hyphen.tex

\def\German{\language = 1\relax}
\German
\input germanhyphen.tex
\English
```

12.7.6 The Hyphenation Character, `\hyphenchar`

The character printed as the hyphenation character is normally "-," but a different hyphenation character can be defined for each font: `\hyphenchar\tenrm = '?` would use the question macro as hyphenation character. See the TEXbook for details. If you assign −1 to the hyphenation character of a font, no hyphenation will take place, if this font is active.

12.7.7 Default Hyphen Character, `\defaulthyphenchar`

If no hyphen character has been specified for a font (using the preceding `\hyphenchar` instruction), then the character of which the character code is stored in the `\defaulthyphenchar` parameter will be used. The default setting of this register is obviously to assign the character code of the hyphen:

```
\defaulthyphenchar = '\-
```

12.7.8 \uchyph

Usually, the first character of the part of a hyphened word which *follows* the hyphen is printed lower case. If, on the other hand, you assign a positive value to parameter **\uchyph**, then this character following the hphen will be printer upper case. The default of this register is zero.

12.7.9 Fine Points About the Line Breaking Algorithm

TEX's line breaking algorithm is much more sophisticated than you might think at first. For instance, the impression I have given you so far is that it simply fills up a line as much as possible before it proceeds to the next line. This is wrong.

Without going into every detail of the line breaking algorithm let me make the following important points:

1. The line breaking algorithm does a *global* optimization in that it looks at all line breaks simultaneously. For instance, assume that TEX "tries" a different line break between lines 4 and 5 of a paragraph. Then this might lead to a different line break between lines 5 and 6, and also between lines 6 and 7.
2. In order to ensure top quality typesetting the line breaking algorithm has the notion of *visually compatible* lines. 5.3.5, p. I-131, discusses the terms *tight*, *loose*, etc. The important point is that TEX tries to avoid visually *incompatible* lines. Two lines are visually incompatible if their classifications (loose, right, etc.) are not adjacent.
3. The computations of the line breaking algorithm are all based on *demerits*. Each line is assigned a *demerit* based on some choice of line break points. The goal of the line breaking algorithm is to *minimize* the sum of the demerits of all lines.
4. Besides the demerits computed for each line (see the TEXbook, Chapter 14, for details) the following additional demerits are added. These demerits are responsible for relating consecutive lines:

   (a) `\adjdemerits`. The value of this parameter is added in case two lines are visually incompatible in the sense specified above. The default of this value in the plain format is 10000.

   (b) `\doublehyphendemerits`. The value of this parameter is added if *two consecutive* lines would be hyphened. This demerit is provided, because in general it is undesirable that two consecutive lines are hyphenated in a paragraph. The default for this parameter is 10000 in the plain format.

(c) **\finalhyphendemerits**. Demerits are added, if the last line of a paragraph is hyphenated (regarded undesirable). The default for this demerit is 5000 in the plain format.

(d) **\linepenalty**. The value of this parameter (it's really neither a penalty nor a demerit) is added to the badness of a line, when the demerits of a line are computed. The higher you make this value (the default in the plain format is 10), the more will TeX try to minimize the number of lines in a paragraph.

12.7.10 Penalties Associated with Line Break Computations

The following horizontal penalties are associated with the computation of line breaks. They are *implicit horizontal penalties* which are inserted by TeX into the horizontal list of a paragraph.

1. **\hyphenpenalty**. Penalty associated with a line break point in a hyphenated word, if the pre-break text is non-empty. This penalty is 50 in the plain format.
2. **\exhyphenpenalty**. Same as **\hyphenpenalty** in case the pre-break text is empty. This penalty is set to 50 in the plain format.
3. **\binoppenalty**. Penalty associated with a line break after a binary operation (inside an inline mathematical equation). The default of this penalty in the plain format is 700.
4. **\relpenalty**. Penalty associated with a line break at a math relation operator. The value of this penalty is 500 in the plain format.

Note that in addition to the above penalties you can insert an explicit penalty into the horizontal list of a paragraph using **\penalty**.

The list of *vertical* penalties in the context of paragraph typesetting can be found in 32.5.4.2, p. IV-20.

12.7.11 Turning Off Hyphenation

There are several different ways to instruct TeX not to hyphen:

1. Assign a value of 10000 to **\hyphenpenalty**. This suppresses hyphenation regardless of the font used, until the value is changed again.
2. Change the hyphenation character for the font, for which you want to suppress hyphenation, to -1. Here is an example: `\hyphenchar\tenrm = -1`

suppresses the hyphenation for **\tenrm**. The hyphenation of text set in other fonts is not affected.

3. Change the hyphenation of a word with the **\hyphenation** command. If you don't use a hyphenation in this word, the word will not be hyphenated.
4. A word inclosed in an hbox will not be hyphened.
5. If you set **\pretolerance** to 10000, pass 2 will not be started and no hyphenation will take place.
6. If the sum of **\lefthyphenmin** and **\righthyphenmin** is 63 or more, no hyphenation takes place.

12.7.12 Hyphenation of Compound Words, \hyph, \slash

By default TEX does not hyphenate compound words, because the hyphen in the compound word and the hyphen resulting from a hyphenation could be confusing to the reader. For that reason not hyphening compound words is a good typesetting practice.

Macro **\hyph** can be used instead of hyphens in compound words allowing TEX to hyphenate at those locations. TEX will hyphen at the hyphen, if this should be necessary. Additionally hyphenation of the component words of a compound word is allowed too. In essence TEX then looks at a compound word on a component by component basis.

The macro **\hyph** is defined as follows:

```
\def\hyph{-\penalty0\hskip0pt\relax}
```

The **\penalty0** tells TEX that there is a neutral break point. **\hskip 0pt** (where a break might occur) tricks TEX into interpreting the components of a compound word as separate words. The **\relax** separates the glue specification from a following text which potentially could start with keywords **plus** or **minus**; see 27.4, p. III-431.

You may find it useful to define macro **\slash** in a similar way replacing "-" by "/".

12.7.13 Print Hyphenation of a Word, \PrintHyphens

The macro **\PrintHyphens** indicates where a word is hyphenated. This macro has one argument, **#1**, which is a word. This word will be printed and those places where TEX might hyphenate this word, are indicated by a hyphen. The main difference between this macro and **\showhyphens** is that **\PrintHyphens** does not write its result to the log file, but prints the word in a hyphenated form, in its word parts. A *word part* of a word is an individually hyphenated part of the word. For instance, the word parts of *mathematics* are *math*, *e*, *mat* and *ics*.

$\mathcal{P}'$ • **prhyph.tip** •

```
\InputD{shboxes.tip}                    % 4.5.15, p. I-111.
\def\PrintHyphens #1{%
```

Start a group.

```
    {%
```

In box register 0 build the hyphenated word with one word part per line. The word is hyphenated at *every possible* hyphenation point.

```
        \setbox 0 = \vbox{%
```

Go to pass 2 of the line breaking algorithm right away.

```
            \pretolerance = -1
```

Force a line break after every hyphen.

```
            \hyphenpenalty = -10000
```

A zero line length means that every word, regardless of how short it is, is "too long."

```
            \hsize = 0pt
            \leftskip = 0pt
            \rightskip = 0pt
            \parfillskip = 0pt
            \parindent = 0pt
```

Don't complain about overfull hbxoxes.

```
            \hfuzz = \maxdimen
```

By setting all implicit vertical penalties inserted by TEX to zero, none of these penalties are inserted in the first place (see 32.5.5, item 6, p. IV-25). It is therefore *not* necessary to use `\unpenalty` (8.5, p. I-301) to remove those penalties from the vertical list of box register 0.

```
            \interlinepenalty = 0
            \clubpenalty = 0
            \widowpenalty = 0
            \brokenpenalty = 0
```

Now print the word, one part of it per line. The `\hskip` preceding the word has the following purpose: TEX only hyphenates words following a glue item (outside of math formulas, of course). By inserting the following `\hskip` the first word is preceded by glue and can and will be hyphenated (see the TEXbook, page 454, second paragraph, for details).

```
            \hskip 0pt
            #1
        }%
```

In box register 2 collect the word parts. This register needs to be initialized to an empty box. The word parts are collected by picking up one line at a time from box register 0, because each such line contains precisely one word part.

```
        \setbox2 = \hbox{}%
```

The vbox just built contains all the word fragments on a line by line basis. Now break that vbox up.

```
35            \setbox 9 = \vbox{%
```

Write out the vertical list of the vbox, in which you collected the word fragments.

```
36                \unvbox 0
```

Start a loop.

```
37                \loop
```

Remove any interline glue, if there is such. If there is none, then `\unskip` has no effect. No interline glue is true once the very last word part was picked up.

```
38                    \unskip
```

Get the word part (`\lastbox`) furthest on the bottom. This word part is removed from the vertical list, so all word fragments are collected backwards. Note that `\lastbox` will be void, if nothing is left on the vertical list in box 0, that is if no word parts are left.

```
39                    \setbox 1 = \lastbox
```

The following test is true *only* if a word part was still left on the vertical list of box register 0, and it was assigned to box register 1.

```
40                    \ifhbox 1
```

Combine the preceding word part with the current word part, insert a discretionary hyphen so TEX can actually hyphenate there later, if necessary and continue to iterate.

```
41                        \global\setbox 2 = \hbox{%
42                            \unhbox 1
43                            \discretionary{}{}{}%
44                            \unhbox 2
45                        }%
46                \repeat
47            }%
```

Come here if the test on line xx was false. Then the word parts are printed out together.

```
48            \unhbox 2
```

End the group this macro was working in.

```
49        }%
50    }
```

• End of `prhyph.tip` •

Here is an example application of this macro. The following input was used.

```
1  \input inputd.tip
2  \InputD{prhyph.tip}                    % 12.7.13, p. 146.
3      \PrintHyphens{Exemplification}
4      \PrintHyphens{of}
5      \PrintHyphens{difficult}
```

```
    \PrintHyphens{and}
    \PrintHyphens{obnoxiously}
    \PrintHyphens{tricky}
    \PrintHyphens{exemplification}
    \PrintHyphens{examples}
    \PrintHyphens{makes}
    \PrintHyphens{me}
    \PrintHyphens{feeling}
    \PrintHyphens{rather}
    \PrintHyphens{funny}.
```

The output by the preceding code reads as follows (ignore the fact for now that TEX does not find all the hyphens):

Ex-em-pli-fi-ca-tion of dif-fi-cult and ob-nox-iously tricky
ex-em-pli-fi-ca-tion ex-am-ples makes me feel-ing rather funny.

12.7.14 Hyphenation, This Series

For the hyphenation control of this series the following file is loaded.

$\mathcal{P}'$ • `ts-hyph.tip` •

the setting of the following two parameters is already part of the plain format which comes with TEX 3.0 and is repeated here for clarity only.

```
\lefthyphenmin  = 2
\righthyphenmin = 3
```

Furthermore note that some other hyphenation problems in this series were corrected using discretionary hyphens.

```
\hyphenation{Ado-be}
\hyphenation{after}
\hyphenation{base-line-skip}
\hyphenation{man-u-script}
\hyphenation{obey-lines}
\hyphenation{obey-spaces}
\hyphenation{other-wise}
```

• End of `ts-hyph.tip` •

12.7.15 Demerits and Additional Parameters in the Typesetting of a Paragraph

The linebreaking algorithm is quite complex, and we will not discuss all the details here; see the TEXbook for further information. Nevertheless a brief overview over those parameters is given here.

There are some additional parameters and *demerits* which control the typesetting of a paragraph. A demerit is something like a penalty, but a demerit's value cannot be computed in advance and depends on the particular line break chosen. Demerits are in units of "badness squared" so you should not be surprised about the rather large values specified below for defaults.

TeX has the following demerits:

- `\adjdemerits`. This demerit is added in case two adjacent lines are visually incompatible. The term *visually incompatible* is explained in 12.7.9, p. 143.
- `\doublehyphendemerits`. This demerit is added if two consecutive lines end with discretionary breaks.
- `\finalhyphendemerits`. This demerit is added if the second to the last line ends with a discretionary break.

12.8 Various Topics

Here near the end of this chapter we will discuss some topics that don't seem to fit in anywhere else.

12.8.1 `\looseness`

The user has the possibility to specify with the help of `\looseness` the number of lines the paragraph should have beyond that the paragraph would have normally. Note though that the attempt to make the paragraph longer or shorter by the number of lines specified is *only* made if the required tolerances (`\tolerance` and `\pretolerance`) do not need to be exceeded.

Here is an example. The following input was used to generate Fig. 12.7, p. 151. Note that `\spaceskip` and `\xspaceskip` were set to rather large values so TeX was able to adjust the interline glue. If those changes would not have been done, the output from the last two examples, for instance, would be the same.

• `looseness.tip` •

```
\def\LoosenessExample #1{
    $$
        \vbox{
            \hsize = 4.0in
            \looseness = #1
            \spaceskip =  12.0pt plus 1in minus 10pt
            \xspaceskip = 12.0pt plus 1in minus 10pt
            This is some paragraph which is set with {\tt
            \string\looseness} = $#1$. This does change the length of
            the paragraph, if there is sufficient glue to stretch or
```

```
                shrink and if the value is a value different than zero.
                We used large values for interword glue (otherwise, for
                instance, in many cases setting {\tt \string\looseness}
                to a negative value will not allow \TeX{} to make
                the paragraph shorter, not even by one line). That's it
                for now.
                \par
            }
      $$
}
```

Sine we have the macro definition, we will now do the following experiments.

```
\LoosenessExample{-2}
\LoosenessExample{-1}
\LoosenessExample{0}
\LoosenessExample{1}
% \LoosenessExample{2}
```

• End of `looseness.tip` •

12.8.2 Typesetting Narrow Columns

Typesetting narrow columns is a real problem: buy any daily newspaper, and see how poor their line breaks sometimes are. The main reason that typesetting narrow columns is such a problem is, of course, the relatively few interword spaces (on a per line basis) one has available for adjusting the line lengths. So these spaces have to be stretched out or pushed together further than in documents with a larger `\hsize`.

Let me first make a couple of suggestions of what you can do to facilitate typesetting of narrow columns. We then will also discuss some additional parameters of TeX which can be used to adjust the line length.

1. Consider using ragged right text.
2. Decrease the penalties and merits associated with hyphenation. You may consider `\pretolerance = -1`, so TeX will immediately try to hyphenate.
3. Increase the stretchability and the stretchability of the interword glue by changing `\spaceskip` and `\xspaceskip` (see 16.2.4, p. 277).
4. Sometimes, though, you have no other choice than to rewrite your text.
5. Don't forget to read the following Subsection.

TeX typically does a very good job. And when it doesn't, then it will also be difficult for you to come up with a better solution.

This is some paragraph which is set with `\looseness` = −2. This does change the length of the paragraph, if there is sufficient glue to stretch or shrink and if the value is a value different than zero. We used large values for interword glue (otherwise, for instance, in many cases setting `\looseness` to a negative value will not allow TEX to make the paragraph shorter, not even by one line). That's it for now.

This is some paragraph which is set with `\looseness` = −1. This does change the length of the paragraph, if there is sufficient glue to stretch or shrink and if the value is a value different than zero. We used large values for interword glue (otherwise, for instance, in many cases setting `\looseness` to a negative value will not allow TEX to make the paragraph shorter, not even by one line). That's it for now.

This is some paragraph which is set with `\looseness` = 0. This does change the length of the paragraph, if there is sufficient glue to stretch or shrink and if the value is a value different than zero. We used large values for interword glue (otherwise, for instance, in many cases setting `\looseness` to a negative value will not allow TEX to make the paragraph shorter, not even by one line). That's it for now.

This is some paragraph which is set with `\looseness` = 1. This does change the length of the paragraph, if there is sufficient glue to stretch or shrink and if the value is a value different than zero. We used large values for interword glue (otherwise, for instance, in many cases setting `\looseness` to a negative value will not allow TEX to make the paragraph shorter, not even by one line). That's it for now.

Figure 12.7. `\looseness` example.

12.8.3 Additional Stretchability, `\emergencystretch`

If people don't care about the quality of their line breaks they might at first try to set `\tolerance = 10000`. The disadvantage with this method is that TeX in this case would have a tendency to consolidate all the badness in one truly horrible line.

If, on the other hand, you set the *dimension register* `\emergencystretch` to a positive value, then the assumption is that additional stretchability of the specified amount is available in the paragraph. The paragraph will therefore be reset with that additional amount.

12.8.4 The Input Form of Paragraphs

Following the philosophy that the input of a TeX document should resemble the output as much as possible, here are some suggestions of how one can make the input more readable:

1. Indicate the beginning of a paragraph by an indentation in the source—after all most likely your `\parindent` is not zero either.
2. The end of a paragraph should be indicated by an empty line, not by `\par`. Even if the instruction following the end of a paragraph causes an implicit `\par` (like for instance `\bigskip`), leave an empty line after the paragraph.
3. If you changed `\leftskip` or `\rightskip` to quote a paragraph (for instance by using `\narrower`), indent your input too.

12.9 Summary

In this chapter we learned the following:

- The `\parshape` command allows fine control over the typesetting of a paragraph, because it is possible to specify the indentation and length of every line of a paragraph.
- The extended `\XParShape` macro allows one to define the shape of a paragraph divided into line groups, where each line group determines the length of a set of lines.
- There are some parameters such as `\everypar`, `\parindent` and `\parskip`, which must be set to their proper values *before* the current paragraph is started. Any changes to these values afterwards do not affect the current paragraph. Then there are parameters, such as `\baselineskip`, `\leftskip` and `\rightskip`, for which the values valid at the very end of a paragraph apply to the whole paragraph (regardless of how often these values were changed in the middle of a paragraph).
- `\vadjust` can be used to insert material into the vertical list (outside of typesetting a paragraph), to which the paragraph's lines will be contributed later.

- A generic table of contents typesetting macro `\GenTocEntry` was presented. This macro is usually called from a macro such as `\TocEntry` or `\EntroIntotoc`, as it would be found in a list of figures file. The macro `\GenTocEntry` can also be used to generate lists of figures or lists of tables.
- The parameter `\prevgraf` contains the number of lines a paragraph has after the paragraph has been typeset. This register can also be set to set a specific starting line number when typesetting a paragraph.
- A macro `\FirstLineSpecial` was presented which typesets the first line of a paragraph in a different font then the rest of the paragraph.
- TeX's line breaking algorithm uses two passes. The first pass does not use hyphenation and succeeds, if no line's badness exceeds `\tolerance`. If it is not possible to find such line breaks, hyphenation is tried, although this time the crucial threshold value is the one loaded in `\pretolerance`.
- Hyphenation in TeX is multilingual. Which language is regarded as active depends on the value stored in `\language`. Hyphenation patterns can only be loaded when `initex` is executed.
- Exceptions to the hyphenation rules can be programmed with discretionary hyphens or the `\hyphenation` primitive can be used to alter the hyphenation of a word for a whole document.
- There are a variety of ways to suppress hyphenation, one of them is to enclose a word inside an hbox.
- The macro `\showhyphens` displays the hyphenation points, as they would be chosen by TeX, in the log file. `\PrintHyphens` can be used to print the same information in some text.
- `\looseness` can be used to make a paragraph artificially shorter or longer, if this is possible without exceeding any of the other existing threshold values.

13 Typesetting Math Equations with TEX

This chapter discusses the typesetting of mathematical formulas with TEX. When writing mathematical formulas the user gives a description of the formula, which resembles very much the way this formula is interpreted mathematically. Actually, this is the way TEX proceeds typesetting mathematical formulas: first the formula's structure is reconstructed from the user's input (by structure we mean the mathematical meaning of the formula). This structure is similar to an expression tree in a compiler. This expression tree is subsequently used to typeset the formula.

All *serious* mathematicians will have to suffer in the two math chapters, because of the liberty I took of using fantasy formulas in most cases.

You will find the presentation of the math mode rather terse at some points. The reason for this is quite simply that typesetting mathematical equations is really fairly straightforward once you get into it. Simply try it and you'll find out how easy it is.

13.1 Basics of Typesetting Mathematical Equations

First we will deal with the very basics of the typesetting of mathematical equations.

13.1.1 Math Mode and Display Math Mode

There are two different math modes in TEX:

1. The first math mode is simply called *math mode* and it is used to write *inline* equations such as $a_{i,j} + a_{j,l} = 0$. To produce this equation you would have to enter `$a_{i,j} + a_{j,l} = 0$`. Observe the simple `$` signs surrounding the formula. Sometimes, to make a clear distinction between this math mode

(for inline equations) and display math mode (explained below), we will call this math mode *inline math mode.* This is *not* a term used in the TEXbook.

2. The second way of typesetting mathematical equations is called *display math mode.* An equation in display math mode is printed *centered*, on a *separate line* and set off from the preceding and the following text by some small extra vertical space.

   For instance, if you enter **`$$a_{i,j} + a_{j,l} = 0$$`**, the output reads as follows (note the double dollar signs surrounding the formula):

$$a_{i,j} + a_{j,l} = 0$$

13.1.2 General Rules for the Math Modes

The following rules apply to both math modes:

1. In general you can assume that anything which works in math mode also works in display math mode and vice versa. Differences will be explained when necessary.
2. The output generated by inline mode differs from display mode not only with respect to *where* the equation is printed, but also the equations will come out differently. These differences are minor and will be explained later. For instance, the size of various symbols (such as the summation sign) is different in math mode and display math mode. For instance, the formula $\sum_{i=1}^{n} x_i$ when printed in display math mode prints as follows:

$$\sum_{i=1}^{n} x_i$$

3. Spaces in the input are ignored. The math modes do their own spacing. Of course the user is able to insert spaces to change the spacing as it is done by TEX (this is rarely necessary). See 13.7, p. 184, for details.

   Spaces can be still syntactically significant: "**`$\sum x$`**" is certainly different from "**`$\sumx$`**."

   The fact that spaces are ignored, allows the user to type in a formula in any format he or she wants.
4. The *math shift character* "**$**" can be printed in regular text by simply escaping it, by writing "**\$**."
5. The reason that the dollar sign acts as a math shift character is that category code 3 has been assigned to it; see 18.1.6, p. III-5, for details on category codes.
6. The "**$**" and "**$$**" which begin and end the math modes also act as group delimiters so that math formula typesetting happens inside an implicit group. See 19.4.10, p. III-106, for details on implicit groups.

7. After the opening "**$**" or "**$$**" are read, the token lists stored in the token registers **\everymath** or **\everydisplay** are read in; see 14.11, p. 223, for details.
8. The curly braces have an additional function as delimiters in the case of fractions delimiting subformulas (see 13.3.2, p. 172) in the math modes.
9. By default regular letters are printed in italics in the math modes, just the way mathematicians like it. For instance, the input **$A-B$** prints $A - B$.
10. A **\par** token (or for that matter a blank line) is *not* allowed in any of the math modes. See the following Subsection for details.

13.1.3 Error Confinement: No Empty Lines in Math Formulas

No empty lines in math formulas more accurately expressed means, of course, no **\par**s in math formulas. What's the reason for this?

Let's assume you make an error omitting a closing math delimiter (right after the formula `x^{2} - 1`), as shown in the following example (the **\hfil\break** was inserted into the example to prevent the output from running into the right margin of the text):

• **ex-math-par.tip** •

```
\nonstopmode

    To prove that $x^{2} - 1 has two zeros $-1$ and $1$, just plug
in the\hfil\break two values and evaluate the formula.

    Now let's look at $x^{2} + 1$. This polynomial cannot be
solved relying on real numbers only.
\bye
```

• End of `ex-math-par.tip` •

First let us discuss what TEX does, then let us discuss why TEX was designed that way. Here is what TEX does: the output generated by the preceding source file looks as follows:

> To prove that $x^2 - 1 has two zeros$-1and1, just plug in the two values and evaluate the formula.$
>
> Now let's look at x^2+1. This polynomial cannot be solved relying on real numbers only.

The preceding output generates the following log file:

• **ex-math-par.log** •

```
This is TeX, C Version 3.14 (...)
**&/usr/local/tex/lib/fmt/plain ex-math-par.tip
(ex-math-par.tip
! Missing $ inserted.
<inserted text>
$
<to be read again>
```

```
\par
1.5

I've inserted a begin-math/end-math symbol since I think
you left one out. Proceed, with fingers crossed.

[1] )
Output written on ex-math-par.dvi (1 page, 636 bytes).
```

What happens is obvious: because a closing math delimiter is missing, TEX takes the rest of the paragraph as a math formula and generates some awful looking output. When TEX sees the end of the paragraph, though, it says to itself "this can't be true: a paragraph ended, but math mode is still in effect. Let me terminate the math mode."

Not only is this a very reasonable thing to do, but imagine TEX would *not* take that corrective action and end your formula. The opening math delimiter of the *following* formula would then be interpreted as the *closing* math delimiter of the current formula. In other words, because of a minor error, TEX, for the rest of the document, would interpret what was intended to be part of a math formula as regular text, and what was intended to be text, as part of a math formula. See the next Subsection for some additional information.

Empty lines are illegal in inline mode as well as display math mode although each mode prints a different error message.

13.1.4 Beginning and Ending Mathematical Equations

TEX uses the same symbol for beginning and ending mathematical equations either a single or a double dollar sign. As you have seen in the preceding subsections, one of the problems resulting from this is that it is difficult to confine errors.

One way around this is to stop using dollar signs. Instead, macros **\BeginMath** and **\EndMath** can be used. An inline math formula now would be surrounded by **\BeginMath** and **\EndMath** (no arguments), and unless there is an error in your document, those two macros act like simple math delimiters.

All the following macros use conditionals and mode testing. For details see 25.1.23.1, p. III-348.

$\mathcal{P}'$ • mathenv.tip •

```
\def\BeginMath{%
```

If TEX is already in math mode, it is determined whether it is in display math mode (in which case the display math mode is terminated first, math mode is started and an error message is generated) or math mode (in which case an error message is generated).

```
    \ifmath
        \ifinner
            \errmessage{\string\BeginMath: already in math
```

```
                    mode, \string\BeginMath ignored.}%
            \else
                \errmessage{\string\BeginMath: in display math
                    mode, terminated and math mode started.}%
                $$
                $
            \fi
        \else
            $\relax
        \fi
}
```

\EndMath works along the same lines.

```
\def\EndMath{%
```

If TEX is *not* already in math mode, the mode is not changed and an error is generated. Otherwise the mode is changed.

```
        \ifmath
            \ifinner
                $%
            \else
                \errmessage{\string\EndMath: you are in display math
                    mode! Should have used \string\EndDisplayMath!}%
                $$
            \fi
        \else
            \errmessage{\string\EndMath: already in math
                mode, \string\EndMath ignored.}%
        \fi
}
```

Similar macros \BeginDisplayMath and \EndDisplayMath for display math mode can be defined as follows:

```
\def\BeginDisplayMath{%
        \ifmath
            \ifinner
                \errmessage{\string\BeginDisplayMath: in inline
                    math mode, terminate it, start display math.}%
                $
                $$
            \else
                \errmessage{\string\BeginDisplayMath: already in math
                    mode, \string\BeginDisplayMath ignored.}%
            \fi
        \else
            $$
        \fi
}
```

And here is the definition of \EndDisplayMath.

```
\def\EndDisplayMath{%
```

```
    \ifmath
        \ifinner
            \errmessage{\string\EndDisplayMath: inline
                math mode in effect, did you intend to
                write \string\EndMath?}%
            $%
        \else
            $$
        \fi
    \else
        \errmessage{\string\EndDisplayMath: not in math
            mode, \string\EndDisplayMath ignored.}%
    \fi
}
```

• End of `mathenv.tip` •

13.1.5 Superscripts and Subscripts

Superscripts and subscripts were already introduced in 2.7.2, item 6, p. I-18, and item 7, p. I-18. Here is a more detailed discussion.

1. The symbol to begin a superscript is `_`. You can also use macro `\sp`.
2. The symbol to begin a subscript is `^`. You can also use macro `\sb`.
3. These symbols or control sequences are followed by the superscript- or the subscript-formula. See item 5, for a discussion of whether curly braces are needed or not.
4. Superscripts and subscripts are only allowed in math mode. If TEX sees a superscript or subscript otherwise (for instance, inside a paragraph) it will print an error message and enter math mode.
5. The same rules as those for undelimited parameters (21.8.4, p. III-183) of a macro apply to the writing of the superscript or subscript itself:

   - If the superscript or subscript text is only one token long then it *can* follow the `_` or `^` directly, *without* being enclosed in curly braces. Of course, like in the case of macro arguments, it *can* be enclosed in curly braces.
   - If the superscript or subscript is more than one token long, it *must* be enclosed in curly braces.

     Example: if you enter `$x^{2y}$` the output is x^{2y}, whereas if you enter `$x^2y$`, then this is interpreted as `$x^{2}y$` and therefore the output reads x^2y.

   Initially, when you are a novice user, it might be a good practice to always enclose the superscript or subscript inside curly braces.
6. You may have *a superscript and a subscript simultaneously.* If you do, you can specify them in *any order.* So `$a^{2x}_{i,j}$` and `$a_{i,j}^{2x}$`

both generate the same output which is $a_{i,j}^{2x}$. It is recommended, though, that you adopt a specific order to which you adhere.

When you have simultaneous superscripts and subscripts, then they start at the same horizontal position. For instance, in A_2^2 superscript and subscript start at the same position. A superscript is moved in horizontally by a small amount in a case such as P_2^2 (**P^2_2**). If you want superscript and subscript to start at the same position use an empty formula **{}** and attach superscripts and subscripts to that formula as in **$P{}^2_2$** which generates the following output: $P{}^2_2$.

7. You are allowed at most one superscript and one subscript per formula. There are instances where curly braces are necessary to resolve ambiguity. When you enter **a_b_c**, then TEX does not know whether you mean **a_{b_c}** or **${a_b}_c$**. This ambiguity *could* have been resolved by some rule such as left associativity, but it has not. So, the above formula **a_b_c** is illegal and you *must* use curly braces to declare your intentions.
8. Here is a brief example, where the superscript is always specified before the subscript (there are all together 8 different possibilities to specify this particular expression; note the way, for instance, subformula y_b^a is enclosed inside curly braces in the following source code):

```
$$
        x^{y^{a}_{b}}_{z^{c}_{d}}
$$
```

This input generates the following output:

$$x^{y^{a}_{b}}_{z^{c}_{d}}$$

9. Superscripts and subscripts can apply to a single character or to a whole formula or subformula. This formula may be empty (which should, but need not be indicated by "**{}**") as, for instance, **${}_2F_2$** which prints as ${}_2F_2$.
10. Super- and subscript are not only used for what is mathematically a super- or subscript. They are also used in instances where something is placed above or below some formula or operation.

    Here is a brief overview of all the applications of superscripts and subscripts in TEX:

    (a) To indicate *regular super- or subscripts* as in a_i^j (**a_{i}^{j}**).

    (b) *To place items above and below so-called "large operators".* For instance, the input, **$\sum_{i=1}^{n-1}$** (**\sum** prints the summation sign) prints $\sum_{i=1}^{n-1}$, or **$\int_a^b f(x)dx$** which prints $\int_a^b f(x)dx$.

    See 13.5, p. 178, for details on the positioning of these superscripts or subscripts.

    (c) In the context of "*underbracing*" and "*overbracing*" as in

$$\underbrace{x+y+z}_{>0}.$$

The source code reads **$$\underbrace{x+y+z}_{>0}$$** for the preceding example. See 14.6.3, p. 208, for details.

(d) In the context of *standard mathematical operators.* Here is an example: $\lim_{i\rightarrow\infty}$ was generated by `$\lim_{i\rightarrow\infty}$`.

11. `^` is the superscript character, because the category code of this symbol is 7. `_` is the subscript symbol, because the category code of `_` is 8.

    Any other symbols with those category codes could be used for superscripts and subscripts. For details, see Table 18.1, p. III-6.

12. Superscripts and subscripts following a character apply to that character only. If, for instance, through the use of curly braces, a subformula is indicated, then the superscript and subscript are placed according to the sizes of the subformula. Here is an example.

`$((x^2_2)^3_3)^4_4$`	$((x^2_2)^3_3)^4_4$
`${({x^2_2}^3_3)}^4_4$`	${({x^2_2}^3_3)}^4_4$

13. Instead of curly braces you can enclose a superscript or subscript into a pair of `\bgroup` and `\egroup`. For instance, `$x^\bgroup n-1\egroup$` prints x^{n-1}.

14. You can use `\prime` to generate an apostrophe in math mode. For instance, to print

$$y' + y''$$

enter

```
$$
    y^\prime + y^{\prime\prime}
$$
```

15. The preceding source code can be simplified using an apostrophe. The apostrophe is set up as a math active character (see 18.2.9, p. III-23, for details) and compared to using `\prime` the input is simplified as follows:

    - You do not need (in fact you must not) enter a caret for superscript.
    - You do not need (in fact you must not) include the apostrophe or apostrophes inside curly braces.

    Using the apostrophe in the preceding example the new input now reads:

```
$$
    y' + y''
$$
```

This example generates the same output as the preceding formula did where `\prime` was used.

16. See 13.3.7, p. 176, for stacked superscripts and subscripts.

17. Extra space in the amount of `\scriptspace` is added to equations with superscripts or subscripts. See the TEXbook for details.

We close with a collection of examples that also prove that spaces don't matter in TEX's input to mathematical equations:

`$x^{2}$`	x^{2}
`$x^{23}$`	x^{23}
`$x^23$`	x^23
`$x^2 3$`	$x^2 3$
`$x^{2-y}_{y-2}$`	x^{2-y}_{y-2}
`$x_{y-2}^{2-y}$`	x_{y-2}^{2-y}
`$x^{2^{3}}$`	$x^{2^{3}}$
`$x^{2} + y^{3}$`	$x^{2} + y^{3}$
`$x^{2^{3^{4^{5}}}}$`	$x^{2^{3^{4^{5}}}}$
`$y_{x^{2}}$`	$y_{x^{2}}$
`$y_{x_{2}}$`	$y_{x_{2}}$

13.1.6 Error Confinement: Superscript and Subscript Related

Similar to the way TEX closes a math formula if an empty line is found, TEX has a mechanism dealing with superscripts and subscripts that are not placed properly. Let's say you enter the following erroneous source code:

• `ex-math-sup.tip` •

```
\nonstopmode

    To prove that x^{2} - 1 has two zeros $-1$ and $1$, note ...
\bye
```

• End of `ex-math-sup.tip` •

The output generated by the preceding source code reads as follows.

To prove that $\mathrm{x}^2 - 1 hastwozeros{-}1and1, note...$

The preceding output generates the following log file:

• `ex-math-sup.log` •

```
This is TeX, C Version 3.14 (...)
**&/usr/local/tex/lib/fmt/plain ex-math-sup.tip
(ex-math-sup.tip
! Missing $ inserted.
<inserted text>
$
<to be read again>
^
l.3 ^^ITo prove that x^
{2} - 1 has two zeros $-1$ and $1$...
I've inserted a begin-math/end-math symbol since I think
you left one out. Proceed, with fingers crossed.

```

```
! Missing $ inserted.
<inserted text>
$
<to be read again>
\par
\bye ->\par
\vfill \supereject \end
1.4 \bye

I've inserted a begin-math/end-math symbol since I think
you left one out. Proceed, with fingers crossed.

[1] )
Output written on ex-math-sup.dvi (1 page, 424 bytes).
```

TEX forces math mode, when it sees the superscript without math mode. Subsequently TEX switches back and forth between math mode and text mode in a way which was not intended (well, if TEX could read your mind, it wouldn't even report the error but silently correct it for you), but eventually it will recover correctly and proceed properly.

13.2 Special Math Symbols and Characters

There are many special symbols and different alphabets available in math mode. We will summarize them here briefly, but first we will look at how TEX's math mode works internally.

13.2.1 The Processing of Mathematical Equations in TEX

It was already discussed that spaces in the input to mathematical equations are ignored, and if you read 13.7, p. 184, then you will find out, that while there are a few instances where the user has to intervene with the horizontal spacing in mathematical equations, these cases are quite rare. You also have seen already that TEX determines the sizes of superscripts (and superscripts of superscripts), etc., properly. The question that arises now is how all of this works.

To discuss all the details is beyond this series, in particular, because it usually works too well, and therefore a user frequently simply does not care. Still, for the rest of this and the following chapter on math mode, it is very useful to have a rough idea of what happens.

The basic principle of TEX is to take apart a math formula like an ordinary compiler does, build its syntax tree (to find out its mathematical meaning), and then based on that compute the sizes of operands, the horizontal spacing and so forth. You can actually observe TEX's way of taking a formula apart using `\showlists`; see 14.1.1, p. 189, for details.

While a formula is taken apart, TEX among many other things classifies operations. The most important classifications of operations are:

1. Binary operations; see 13.2.7, p. 167.
2. Relational operations; see 13.2.8, p. 168.
3. Operations of type Ord; see 13.2.6, p. 167.

For instance, if you look at the following formula

$$a * b < c * d$$

then you will realize, if you look closely, that the space around the less than sign (a relational operation) is slightly larger than the space around the multiplication sign (binary operations). This is done by TEX to emphasize the precedence relations of these operations. The above formula, written with parenthesis, is interpreted as

$$(a * b) < (c * d)$$

and not as

$$a * (b < c) * d$$

13.2.2 Variables in Math Equations

As you probably realized by now, variables in math mode are set using italics. This is what you usually want to happen. On the other hand, $lim_{i\to\infty} x_i$ is wrong and should read $\lim_{i\to\infty}$ instead.

You can achieve roman characters in math mode using `\rm`. You will need this rarely, because the plain format defines a set of macros which do all of this and in addition classify the used symbols as operations (`\mathop`) so that the spacing works out properly. The following sequences are defined (the output generated by these symbols is not reprinted, because it is obvious: `$\lim$` prints "lim" and so forth).

`\arccos`	`\arcsin`	`\arctan`	`\arg`	`\cos`	`\cosh`
`\cot`	`\coth`	`\csc`	`\deg`	`\det`	`\dim`
`\exp`	`\gcd`	`\hom`	`\inf`	`\ker`	`\lg`
`\lim`	`\liminf`	`\max`	`\min`	`\Pr`	`\sec`
`\sin`	`\sinh`	`\sup`	`\tan`	`\tanh`	

Be careful not to mix up the number 0 and the letter O.

Table 13.1. Table of lowercase Greek letters accessible in math mode.

\alpha	α	\beta	β	\gamma	γ
\delta	δ	\epsilon	ϵ	\zeta	ζ
\eta	η	\theta	θ	\iota	ι
\kappa	κ	\lambda	λ	\mu	μ
\nu	ν	\xi	ξ	o	o
\pi	π	\rho	ρ	\sigma	σ
\tau	τ	\upsilon	υ	\phi	ϕ
\chi	χ	\psi	ψ	\omega	ω

Table 13.2. Table of uppercase Greek letters accessible in math mode.

\Gamma	Γ	\Delta	Δ
\Theta	Θ	\Lambda	Λ
\Xi	Ξ	\Pi	Π
\Sigma	Σ	\Upsilon	Υ
\Phi	Φ	\Psi	Ψ
\Omega	Ω		

13.2.3 Lowercase Greek Letters

To get a lowercase Greek letter you simply write the name of the Greek letter: **α** to get α, **\beta** to get β, and so forth. A complete table of the available lowercase Greek letters can be found in Table 13.1 on this page.

Some of the lowercase Greek letters have variants. Here is a table printing the two variants of four different Greek letters side by side.

\phi:	ϕ	\varphi:	φ
\theta:	θ	\vartheta:	ϑ
\epsilon:	ϵ	\varepsilon:	ε
\rho:	ρ	\varrho:	ϱ

The symbol for the empty set $\emptyset$ (**$\emptyset$**) is different from the Greek letter ϕ (**ϕ**). Also the Greek letter ϵ (**ϵ**) is different from the symbol $\in$ (**$\in$**) used to denote membership in a set.

While we are at discussing Greek characters, here is how partial derivatives are written: to typeset $\frac{\partial}{\partial x}$ enter **$\partial \over \partial x$**. The use of **\over** to generate a fraction will be discussed shortly.

13.2.4 Uppercase Greek Letters

To make capital Greek letters you write the name of the letter with an initial capital letter; thus writing **Γ** prints Γ. However, since most of the uppercase Greek letters are identical to uppercase letters of the English alphabet, no control sequences with their Greek name exists for most of the Greek letters.

Table 13.3. Symbols of type Ord.

`\aleph:`	$\aleph$	`\hbar:`	$\hbar$
`\imath:`	$\imath$	`\jmath:`	$\jmath$
`\ell:`	ℓ	`\wp:`	$\wp$
`\Re:`	$\Re$	`\Im:`	$\Im$
`\partial:`	∂	`\infty:`	∞
`\prime:`	$\prime$	`\emptyset:`	$\emptyset$
`\nabla:`	∇	`\surd:`	$\surd$
`\top:`	$\top$	`\bot:`	$\bot$
`|`	$\|\bot$	`\angle:`	$\angle$
`\triangle:`	$\triangle$	`\backslash:`	$\backslash$
`\forall:`	$\forall$	`\exists:`	$\exists$
`\neg:`	$\neg$	`\flat:`	$\flat$
`\natural:`	$\natural$	`\sharp:`	$\sharp$
`\clubsuit:`	$\clubsuit$	`\diamondsuit:`	$\diamondsuit$
`\heartsuit:`	$\heartsuit$	`\spadesuit:`	$\spadesuit$

13.2.5 Other Alphabets

There is a *calligraphic* alphabet (capital letters only) which can be selected with `\cal`. Thus $\mathcal{A}, \mathcal{B}, \mathcal{C}, \ldots$ has been typeset by writing `$\cal A, B, C, \ldots$`. This alphabet belongs to the text fonts of family 2 (see 14.5.3, p. 203).

There is also a *math italic* alphabet which is selected by `\mit`. It prints uppercase Greek letters like $\mathit{\Gamma}$ instead of the usual Γ. This font in general has somewhat wider characters than regular text italics. In particular it has a wide f, which is desirable for writing a function as in $f(x)$. This is most obvious when writing a word such as "different" which has two adjacent "f"s. Using the `\mit` font this comes out as $different$, using text italics it prints "*different*." Obviously you would never write this word using `\mit`.

13.2.6 Miscellaneous Symbols of Type Ord

Miscellaneous symbols of type Ord are presented in Table 13.3 on this page.

13.2.7 Binary Operations (Type Bin)

TEX requires no special control sequences for binary operations $+$ and $-$. There is a wealth of additional binary operations displayed in Table 13.4 on the next page.

Table 13.4. Binary operations.

`+`	$+$	`-`	$-$
`\pm:`	$\pm$	`\mp:`	$\mp$
`\setminus:`	$\setminus$	`\cdot:`	$\cdot$
`\times:`	$\times$	`\ast:`	$\ast$
`\star:`	$\star$	`\diamond:`	$\diamond$
`\circ:`	$\circ$	`\bullet:`	$\bullet$
`\div:`	$\div$	`\cap:`	$\cap$
`\cup:`	$\cup$	`\uplus:`	$\uplus$
`\sqcap:`	$\sqcap$	`\sqcup:`	$\sqcup$
`\triangleleft:`	$\triangleleft$	`\triangleright:`	$\triangleright$
`\wr:`	$\wr$	`\bigcirc:`	$\bigcirc$
`\bigtriangleup:`	$\bigtriangleup$	`\bigtriangledown:`	$\bigtriangledown$
`\vee:`	$\vee$	`\wedge:`	$\wedge$
`\oplus:`	$\oplus$	`\ominus:`	$\ominus$
`\otimes:`	$\otimes$	`\oslash:`	$\oslash$
`\odot:`	$\odot$	`\dagger:`	$\dagger$
`\ddagger:`	$\ddagger$	`\amalg:`	$\amalg$

13.2.8 Relational Operations (Type Rel)

There are plenty of relational operations available in TEX, as Table 13.5 on the next page shows. Many of the relations can be negated easily: simply write `\not` in front of the operation. For instance, `$A\subset B$` prints $A \subset B$, while `$A\not\subset B$` prints $A \not\subset B$. The less than and greater than signs $<$ and $>$ also belong to relational operators. Instead of `\geq` you can also write `\ge`.

The following macro is used to print syntactic variables in programming language theory. For instance, the input `\angt{expr}` prints ⟨expr⟩. This macro has one argument, `#1`, the text to be enclosed in angle brackets.

$\mathcal{P}'$ • `angt.tip` •

```
\def\angt #1{%
```

Switch to horizontal mode, if necessary, so `\angt{...}` can appear at the beginning of a paragraph.

```
    \leavevmode
```

Note: you may already be in math mode, so to be safe include the following "mathematical equation" (consisting of a "greater than" sign only) into an hbox.

```
    \hbox{$\langle$}%
```

Print the text itself and then the ">."

```
    {\rm #1}%
    \hbox{$\rangle$}%
}
```

• End of `angt.tip` •

Table 13.5. Relational operations.

`>`	$>$	`<`	$<$
`\leq:`	$\leq$	`\prec:`	$\prec$
`\preceq:`	$\preceq$	`\ll:`	$\ll$
`\subset:`	$\subset$	`\subseteq:`	$\subseteq$
`\sqsubseteq:`	$\sqsubseteq$	`\in:`	$\in$
`\vdash:`	$\vdash$	`\smile:`	$\smile$
`\frown:`	$\frown$	`\geq:`	$\geq$
`\succ:`	$\succ$	`\succeq:`	$\succeq$
`\gg:`	$\gg$	`\supset:`	$\supset$
`\supseteq:`	$\supseteq$	`\sqsupseteq:`	$\sqsupseteq$
`\ni:`	$\ni$	`\dashv:`	$\dashv$
`\mid:`	$\mid$	`\parallel:`	$\parallel$
`\equiv:`	$\equiv$	`\sim:`	$\sim$
`\simeq:`	$\simeq$	`\asymp:`	$\asymp$
`\approx:`	$\approx$	`\cong:`	$\cong$
`\bowtie:`	$\bowtie$	`\propto:`	$\propto$
`\models:`	$\models$	`\doteq:`	$\doteq$
`\perp:`	$\perp$		

Here is a brief example application of the preceding macro.

```
\angt{expr}, \angt{expr$_{2}$},
$X + \angt{QQ} + \angt{ZZ}$, and other
things we can do.
```

The preceding text generates the following output:

$\langle$expr$\rangle$, $\langle$expr$_2\rangle$, $X + \langle QQ\rangle + \langle ZZ\rangle$, and other things we can do.

13.2.9 Arrows

There are a number of horizontal and vertical arrows. Those are include in Table 13.6 on the next page. Note that some of the vertical arrows can be also used as large delimiters (see 13.6, p. 180).

13.2.10 Special Symbols

Here is a list of just a few of these special math symbols.

1. *Roots*, the square root symbol. To print $\sqrt{x+2}$ `\rootwrite $\sqrt {x+2}$`. For arbitrary roots, use `$\root n+1\of {x^2+y^2}$` to print $\sqrt[n+1]{x^2+y^2}$.

Table 13.6. Arrows.

`\leftarrow:`	$\leftarrow$	`\Leftarrow:`	$\Leftarrow$
`\rightarrow:`	$\rightarrow$	`\Rightarrow:`	$\Rightarrow$
`\leftrightarrow:`	$\leftrightarrow$	`\Leftrightarrow:`	$\Leftrightarrow$
`\mapsto:`	$\mapsto$	`\hookleftarrow:`	$\hookleftarrow$
`\leftharpoonup:`	$\leftharpoonup$	`\leftharpoondown:`	$\leftharpoondown$
`\rightleftharpoons:`	$\rightleftharpoons$	`\longleftarrow:`	$\longleftarrow$
`\Longleftarrow:`	$\Longleftarrow$	`\longrightarrow:`	$\longrightarrow$
`\Longrightarrow:`	$\Longrightarrow$	`\longleftrightarrow:`	$\longleftrightarrow$
`\Longleftrightarrow:`	$\Longleftrightarrow$	`\longmapsto:`	$\longmapsto$
`\hookrightarrow:`	$\hookrightarrow$	`\rightharpoonup:`	$\rightharpoonup$
`\rightharpoondown:`	$\rightharpoondown$	`\uparrow:`	$\uparrow$
`\Uparrow:`	$\Uparrow$	`\downarrow:`	$\downarrow$
`\Downarrow:`	$\Downarrow$	`\updownarrow:`	$\updownarrow$
`\Updownarrow:`	$\Updownarrow$	`\nearrow:`	$\nearrow$
`\searrow:`	$\searrow$	`\swarrow:`	$\swarrow$
`\nwarrow:`	$\nwarrow$		

2. TEX has two different *l*s. The regular l in math mode is typeset by writing **`$l$`**, while there is also ℓ which you get by writing **`$\ell$`**.
3. *Dotless i and dotless j*: sometimes (for instance, when using an accent), you want a dotless i and j—you get this by **`$\imath$`** and **`$\jmath$`** which prints $\imath$ and $\jmath$.
4. As already mentioned, there are numerous other *special symbols*, which can be used in math mode. For instance, there is $\approx$ (**`$\approx$`**) and $\mapsto$ (**`$\mapsto$`**).
5. Ellipses can be generated with **`\cdots`** (centered dots) as in **`$1+2+\cdots+n$`** ($1+2+\cdots+n$?), or as in **`$(x_{1}, \ldots, x_{n})$`** ($(x_1,\ldots,x_n)$). See also 2.7.2, item 26.c, p. I-22.

13.2.11 Overline and Underline

The following two control sequences are provided for putting a line above or below some expression:

`$\overline {xyz}$`	$\overline{xyz}$
`$\underline {xyz}$`	$\underline{xyz}$

The macro **`\underline`** works in math mode only and is not very suitable for underlining text. For a detailed discussion of underlining text see 16.7, p. 297.

13.2.12 Text in Math Modes

Especially in display math mode you sometimes want to insert some text into a formula. In inline math mode one can break up a formula by simply exiting inline math mode, writing some text and entering inline math mode again. This is not possible in display math mode. The following approach can be taken in display math mode (unnecessary in inline math mode):

To print the following formula

$$X_n = X_k \qquad \hbox{if and only if} \qquad Y_n = Y_k \quad \hbox{and} \quad Z_n = Z_k$$

the following was entered

```
$$
    X_n=X_k \hbox{\qquad if and only if\qquad}
        Y_n=Y_k \hbox{\quad and\quad} Z_n=Z_k
$$
```

Using hboxes (instead of writing `\rm` to get roman text) has the additional advantage that spaces in math mode are not ignored.

The preceding formula also could have been written as follows. Note that because TEX's math mode ignores spaces, you must insert "\␣s" to have spaces in the final output.

```
$$
    X_n=X_k {\rm\ \ \ if\ and\ only\ if\ \ \ }
        Y_n=Y_k {\rm\ \ and \ \ } Z_n=Z_k.
$$
```

The output generated by the preceding input reads as follows:

$$X_n = X_k \ \ \ \hbox{if and only if} \ \ \ Y_n = Y_k \ \ \hbox{and} \ \ Z_n = Z_k.$$

13.3 Fractions and Fraction-Like Constructs

This Section discusses fractions which are an important ingredient in the typesetting of mathematical equations. We are primarily discussing fractions which are *not* written with a slash ($(a/b)/c$) but with using a fraction bar:

$$\frac{\frac{a}{b}}{c}$$

13.3.1 Fractions with `\over`

Fractions can be typeset using `\over`. This control sequence is applied as follows:

`{` ⟨subformula numerator⟩ `\over` ⟨subformula denominator⟩ `}`

Note that contrary to what you might first assume, the numerator and denominator are *not* enclosed inside curly braces. Instead the numerator is delimited by an opening curly brace and `\over` and the denominator is delimited by `\over` and a closing curly brace.

The following two simplifications are permitted, when it comes to the curly braces of an `\over`.

1. If a fraction forms a whole formula, the `{}` can be omitted. In other words, `${ ... \over ... }$` can be simplified to `$ ... \over ... $`.
2. `\left` and `\right` act as fraction delimiters too. For instance, `$... \left({... \over ...}\right) ...$` can be simplified to `$... \left( ... \over ... \right) ...$`. `\left` and `\right` are discussed in 13.6, p. 180.

A novice user is encouraged to write always fractions with `{}`. Also note that instead of curly braces `\bgroup` and `\egroup` can be used.

13.3.2 `\atop`, `\choose` and `\above`

The following three macros are applied similar to `\over`:

1. `\above` is like `\over` except for the fact that `\above` must be followed by a dimension, the thickness of the fraction bar to be used. For instance, `$a \above 2pt b$` prints $a \above 2pt b$.
2. `\atop` places "numerator" and "denominator" atop of each other without any fraction bar. For instance, `$a-x \atop b$` prints ${a-x \atop b}$.
3. `\choose` is used to write binomial formulas. For instance, `$a\choose b-x$` prints $\binom{a}{b-x}$.

For examples of how the preceding control sequences (including `\over`) are applied see Figure 13.1 on the next page.

13.3.3 General Fractions

TEX knows about the following three types of *generalized fraction primitives*, on top of which the previously introduced fraction macros are built. ⟨delim_1⟩ and ⟨delim_2⟩ are placed to the left and right of the fraction or fraction-like construct.

1. `\abovewidthdelims` ⟨delim_1⟩ ⟨delim_2⟩ ⟨dimen⟩: This is the most general of the generalized fractions. TEX places the immediately preceding subformula

`$${1+a \over b+2}$$`	prints	$\dfrac{1+a}{b+2}$
`$$1+a \over b+2$$`	prints	same, shorter (outermost braces dropped)
`$$1+{a \over b+2}$$`	prints	$1+\dfrac{a}{b+2}$
`$${1+a \over b}+2$$`	prints	$\dfrac{1+a}{b}+2$
`$$1+{a \over b}+2$$`	prints	$1+\dfrac{a}{b}+2$
`$$1+{a \above 1pt b}+2$$`	prints	$1+\dfrac{a}{b}+2$
`$$1+{a \atop b}+2$$`	prints	$1+{a \atop b}+2$
`$$1+{a \choose b}+2$$`	prints	$1+\dbinom{a}{b}+2$
`$$\left({1+a \over b}\right)+2$$`	prints	$\left(\dfrac{1+a}{b}\right)+2$
`$$\left(1+a \over b\right)+2$$`	prints	$\left(\dfrac{1+a}{b}\right)+2$
`$$\left({a \over b}\right)+2$$`	prints	$\left(\dfrac{a}{b}\right)+2$
`$$\left(a \over b\right)+2$$`	prints	$\left(\dfrac{a}{b}\right)+2$
`$$a \over b \over c$$`	is illegal	
`$${a \over {b \over c}}$$`	prints	$\dfrac{a}{\frac{b}{c}}$
`$${{a \over b} \over c}$$`	prints	$\dfrac{\frac{a}{b}}{c}$

Figure 13.1. Fraction examples.

(the numerator) on top of the immediately following subformula (the denominator), with the two separated by a fraction bar of thickness ⟨dimen⟩. **`\above`** ⟨dimen⟩ is an abbreviation of **`\abovewithdelims{}{}`** ⟨dimen⟩.

2. **`\overwithdelims`** ⟨delim_1⟩ ⟨delim_2⟩. This primitive is largely equivalent to **`\abovewidthdelims`**. This time the thickness of the fraction bar is determined by the current font. **`\over`** uses this control sequence, and is defined as **`\overwithdelims{}{}`**.
3. **`\atopwithdelims`** ⟨delim_1⟩ ⟨delim_2⟩. This control sequence is equivalent to **`\abovewidthdelims`** but with no fraction bar. **`\atop`** is defined as **`\atopwithdelims{}{}`**. **`\choose`** is equivalent to **`\atopwithdelims()`**.

13.3.4 Entering Fractions

The most commonly made "error" when inputting fractions is to enclose the numerator and denominator inside curly braces separately, in other words to write

```
{num} \over {den}
```

instead of

```
{num \over den}
```

The word error was quoted, because you might actually want curly braces such as when the numerator or denominator are fractions themselves. Also note that if you have a formula written as **${a} \over {b}$**, then those curly braces, while not necessary, do not hurt either.

To keep track of curly braces when entering fractions is difficult. Here are some suggestions for how to avoid problems:

1. Complete the fraction construct first using dummy numerators and denominators but include the necessary curly braces right from the beginning.
2. Fill in numerator and denominator later.
3. Follow strict indentation rules, in particular when the fraction is complicated: **{**, **\over** and **}** should be indented on the same level. Numerator and denominator should be indented by one additional level.
4. You might want to add a horizontal line (as a comment) after **\over** in the input, to indicate the fraction bar.

Here is a concrete example. Assume that you like to typeset the following formula:

$$\frac{\sum_i \frac{a_i}{b_i}}{\sum_j \frac{d_j}{e_j}}$$

Here are the steps you should go through:

1. Enter an outline of the outermost fraction.

```
{
    numerator (a sum)
\over % -------------
    denominator (a sum)
}
```

2. Fill in the numerator slowly.

```
{
    \sum_i {
                numerator
            \over % --------------
                denominator
            }
\over % -------------
    denominator
}
```

3. The numerator is expanded further.

```
{
    \sum_i {
                a_i
```

```
                   \over % ---------------
                       b_i
                   }
        \over % --------------
            denominator (a sum)
        }
```

4. And then do the same to the denominator to get what you wanted.

```
        {
            \sum_i {
                       a_i
                   \over
                       b_i
                   }
        \over
            \sum_j {
                       d_j
                   \over
                       e_j
                   }
        }
```

For a small fraction, like $\frac{a_i}{b_i}$, you probably will write `{a_i\over b_i}` instead of

```
            {
                a_i
            \over % ----------
                b_i
            }
```

13.3.5 Another Way of Entering Fractions, \frac

You might find it easier to enter fractions by using the `\frac` macro[1] with the numerator and denominator being the two parameters `#1` and `#2` of this macro. Thus the fraction

$$\frac{\frac{a}{b}}{c}$$

can be entered as `$$\frac{\frac{a}{b}}{c}$$` using this macro. Here is the definition of the `\frac` macro.

$\mathcal{P}'$ • frac.tip •

```
\def\frac #1#2{%
    {#1 \over #2}%
}
```

[1] This macro is, by the way, the only way fractions are introduced in LaTeX. You still can use `\over` in fractions in LaTeX-based documents, of course.

• End of `frac.tip` •

13.3.6 Using a Slash Instead of Fractions

Instead of writing `\over` you should always keep the use of a slash in mind. For instance,

$$\binom{n}{k/2}$$

looks much better than

$$\binom{n}{\frac{k}{2}}$$

13.3.7 Stacked Superscripts and Subscripts

Sometimes you need stacked superscripts, as, for instance, in the following formula

$$\sum_{{i=1\atop i\not=m}\atop i {\rm \ is\ not\ prime}}^{n+1} x_{i-1}$$

The input to this formula reads as follows:

```
$$
    \sum_{{i=1\atop i\not=m}\atop i {\rm \ is\ not\ prime}}^{n+1}
                    x_{i-1}
$$
```

If you look at the preceding formula closely, then you will discover that the material above the summation sign and the material below the summation sign are of different sizes. Here is the corrected formula. First the new input (see 14.4.1, p. 195, for an explanation of what is going on).

```
$$
    \sum
    _{
        {
            \scriptstyle i=1
        \atop
            \scriptstyle i \not= m
        }
    \atop
        \scriptstyle i {\rm \ is\ not\ prime}
    }
    ^{n+1}
    x_{i-1}
$$
```

This input generates the following output:

$$\sum_{\substack{i=1\\ i\neq m\\ i \text{ is not prime}}}^{n+1} x_{i-1}$$

13.4 Math Accents

Math accents and text accents are two separate issues in TEX. This Section discusses math accents and wide accents. See 16.5, p. 293, for details on text accents.

13.4.1 Simple Math Accents

Normally accents are above a single character only (remember to use the dotless i and j in these cases). TEX provides 10 different math accents, which are shown next:

Input	Output
`$ \hat a $`	$\hat a$
`$ \check a $`	$\check a$
`$ \tilde a $`	$\tilde a$
`$ \acute a $`	$\acute a$
`$ \grave a $`	$\grave a$
`$ \dot a $`	$\dot a$
`$ \ddot a $`	$\ddot a$
`$ \breve a $`	$\breve a$
`$ \bar a $`	$\bar a$
`$ \vec a $`	$\vec a$

The definitions of these accents use `\mathaccent`.

13.4.2 Wide Accents

Wide accents span more than one character. The following ones are defined in TEX: `\widehat` and `\widetilde`. They come in three different sizes—$\widehat{x}$, $\widehat{xy}$, $\widehat{xyz}$ and $\widetilde{x}$, $\widetilde{xy}$, $\widetilde{xyz}$. They cannot grow any wider; thus `$\widetilde{xyzxyz}$` results in $\widetilde{xyzxyz}$ centered over the middle of the formula.

Table 13.7. Large operators in math mode.

\sum:	$\displaystyle\sum$	$\sum$	\prod:	$\displaystyle\prod$	$\prod$
\coprod:	$\displaystyle\coprod$	$\coprod$	\int:	$\displaystyle\int$	$\int$
\oint:	$\displaystyle\oint$	$\oint$	\bigcap:	$\displaystyle\bigcap$	$\bigcap$
\bigcup:	$\displaystyle\bigcup$	$\bigcup$	\bigsqcup:	$\displaystyle\bigsqcup$	$\bigsqcup$
\bigvee:	$\displaystyle\bigvee$	$\bigvee$	\bigwedge:	$\displaystyle\bigwedge$	$\bigwedge$
\bigodot:	$\displaystyle\bigodot$	$\bigodot$	\bigotimes:	$\displaystyle\bigotimes$	$\bigotimes$
\bigoplus:	$\displaystyle\bigoplus$	$\bigoplus$	\biguplus:	$\displaystyle\biguplus$	$\biguplus$

13.5 Large Operators

Large operators like summation ($\sum$) and integration ($\int$) are symbols which change in appearance depending on whether they are used in inline math mode or display math mode. Only summation and integration is discussed first. We will discuss some specific examples next. A complete list of all large operators can be found in Table 13.7 on this page. This table shows every operator in the two sizes in which it is available.

13.5.1 Summation

Compare a summation in math mode $\sum_{i=1}^{m} x_i$ and one in display math mode:

$$\sum_{i=1}^{m} x_i$$

Note that the following items changed between the two modes:

- The operator is larger in display math mode.
- The limits of the summation in display math mode are placed above and below the summation sign, not to the left and right as in math mode.

13.5.2 Integral Signs

In the case of the *integral sign* the *size* of the operator *changes* (math mode compared to display math mode) but the *positioning of the limits* does *not hange*. Here is an example. Compare $\int_a^b f(x)\,dx$ and

$$\displaystyle\int_a^b f(x)\,dx$$

3.5.3 Changing the Positions of Large Operator-Related ubscripts and Superscripts, `\limits` and `\nolimits`

Vith the control sequence `\limits` the limits are positioned *above* and *below* he operator (like the default for summation) and with the control sequence `nolimits` the limits are positioned to the right of the right upper and lower orners of the symbols (like the default for the integration sign). These control equences must immediately follow the control sequence specifying the large perator. For instance, `$$\sum \nolimits_{n=1}^m$$` results in

$$\sum\nolimits_{n=1}^m$$

he output in display math mode (which is equivalent to using `\limits` instead f `\nolimits`) normally looks as follows:

$$\sum_{n=1}^m$$

3.5.4 Roots and Square Roots, `\sqrt`, `\root`

TEX can typeset even larger symbols. Here is an example. The following input vas used.

```
$$
    \sqrt{1+\sqrt {1+\sqrt{1+\sqrt{1+\sqrt{1+\sqrt{1+\sqrt{1+x}}}}}}}
$$
```

This input generates the following output:

$$\sqrt{1+\sqrt{1+\sqrt{1+\sqrt{1+\sqrt{1+\sqrt{1+\sqrt{1+x}}}}}}}$$

In this case different symbols are selected depending on the required size. The three largest square root signs are essentially the same with the exception of he addition of vertical and horizontal rules of the appropriate length. This last bservation is important: TEX assumes that you can draw vertical and horizontal ules at any position and any size. It is TEX which composes a big root sign out f a "small initial root part ($\sqrt{}$)" sign and a horizontal and vertical rule. In a imilar way delimiters can grow to any desired size.

Of course, you don't need to write something that fancy. `$\sqrt{x}$` prints $\sqrt{x}$. For a general root use `\root`. The "exponent" of a root is enclosed between `\root` and `\of`. The radicand of the root follows `\of`, which, if longer than one token, must be enclosed in curly braces. Here is an example: `$$\root x-1\of {n+1} - 1$$` prints

$$\root x-1\of {n+1} - 1$$

13.6 Delimiters

Delimiters are interesting, because they can grow to any necessary size. The most prominent examples for delimiters are parentheses. For instance, in order to print

$$1 + \left(\frac{1}{1-x^2}\right)$$

the following questions arise:

1. How does TEX select the *size* of the *parenthesis*?
2. How much is this selection *automized* and when is it necessary to *manually* select the size.
3. How is this manual size selection if necessary achieved?

The delimiters provided by TEX are reprinted in Table 13.8 on the next page. We call a *left delimiter* a delimiter appearing on the left side of a delimited formula. The *right delimiter* appears on the right side.

13.6.1 Explicit Specification of the Size of a Delimiter

The user can explicitly specify the size of delimiters as the following example shows (the `\bigl`, `\bigr`, etc., delimiters work with all symbols which can grow to an arbitrary size in TEX):

`$\lbrace$`	prints	$\lbrace$
`$\bigl\lbrace$`	prints	$\bigl\lbrace$
`$\Bigl\lbrace$`	prints	$\Bigl\lbrace$
`$\biggl\lbrace$`	prints	$\biggl\lbrace$
`$\Biggl\lbrace$`	prints	$\Biggl\lbrace$

Use `\bigr`, `\Bigr`, ... for printing *right delimiters*, for instance `\Bigr)` to print a large closing parenthesis.

Table 13.8. Math delimiters.

<table>
<tr><th>Source</th><th>Output</th><th>Output</th></tr>
<tr><td>(</td><td>left parenthesis:</td><td>$($</td></tr>
<tr><td>)</td><td>right parenthesis:</td><td>$)$</td></tr>
<tr><td>[or \lbrack</td><td>right bracket:</td><td>$[$</td></tr>
<tr><td>] or \rbrack</td><td>left bracket:</td><td>$]$</td></tr>
<tr><td>\{ or \lbrace</td><td>left brace:</td><td>$\{$</td></tr>
<tr><td>\} or \rbrace</td><td>right brace:</td><td>$\}$</td></tr>
<tr><td>\lfloor</td><td>left floor bracket:</td><td>$\lfloor$</td></tr>
<tr><td>\rfloor</td><td>right floor bracket:</td><td>$\rfloor$</td></tr>
<tr><td>\lceil</td><td>left ceiling bracket:</td><td>$\lceil$</td></tr>
<tr><td>\rceil</td><td>right ceiling bracket:</td><td>$\rceil$</td></tr>
<tr><td>\langle</td><td>left angle bracket:</td><td>$\langle$</td></tr>
<tr><td>\rangle</td><td>right angle bracket:</td><td>$\rangle$</td></tr>
<tr><td>/</td><td>slash:</td><td>$/$</td></tr>
<tr><td>\backslash</td><td>backslash:</td><td>$\backslash$</td></tr>
<tr><td>| or \vert</td><td>vertical bar:</td><td>$\vert$</td></tr>
<tr><td>\| or \Vert</td><td>double vertical bar:</td><td>$\Vert$</td></tr>
<tr><td>\uparrow</td><td>upward arrow:</td><td>$\uparrow$</td></tr>
<tr><td>\Uparrow</td><td>double upward arrow:</td><td>$\Uparrow$</td></tr>
<tr><td>\downarrow</td><td>downward arrow:</td><td>$\downarrow$</td></tr>
<tr><td>\Downarrow</td><td>double downward arrow:</td><td>$\Downarrow$</td></tr>
<tr><td>\updownarrow</td><td>updownward arrow:</td><td>$\updownarrow$</td></tr>
<tr><td>\Updownarrow</td><td>double updownward arrow:</td><td>$\Updownarrow$</td></tr>
<tr><td></td><td>null delimiter.</td><td></td></tr>
</table>

Then there is `\bigm`, `\Bigm`, etc., which are for *middle delimiters*, for the middle of a formula.

Here is the definition of the preceding macros. These macros use `\left` and `\right` explained in the following Subsection. We begin with a definition of macro `\n@space` which is used in the macros below.

```
\def\n@space{\%
    \nulldelimiterspace = 0pt
    \mathsurround = 0pt
}

\def\bigl{\mathopen\big}  \def\bigm{\mathrel\big}
                          \def\bigr{\mathclose\big}
\def\Bigl{\mathopen\Big}  \def\Bigm{\mathrel\Big}
                          \def\Bigr{\mathclose\Big}
\def\biggl{\mathopen\bigg}\def\biggm{\mathrel\bigg}
                          \def\biggr{\mathclose\bigg}
\def\Biggl{\mathopen\Bigg}\def\Biggm{\mathrel\Bigg}
                          \def\Biggr{\mathclose\Bigg}
\def\big#1{{ \hbox{$\left#1\vbox to  8.5 pt{}\right.\n@space$}}}
\def\Big#1{{ \hbox{$\left#1\vbox to 11.5 pt{}\right.\n@space$}}}
\def\bigg#1{{\hbox{$\left#1\vbox to 14.5 pt{}\right.\n@space$}}}
```

```
\def\Bigg#1{{\hbox{$\left#1\vbox to 17.5 pt{}\right.\n@space$}}}
```

13.6.2 Implicit Specification of the Size of a Delimiter

For implicit specifications of the size of delimiters `\left` and `\right` are used, making the size selection done by TEX. Here is how `\left` and `\right` are applied.

`\left` $\langle \text{delim}_1 \rangle$ $\langle \text{subformula} \rangle$ `\right` $\langle \text{delim}_2 \rangle$

TEX typesets the subformula first to determine its height and depth, and then puts the specified delimiters around it, in a size just big enough to cover the subformula.

The `\left` and `\right` specifiers also have the effect of grouping, so definitions are local unless preceded by `\global`.

Symbols generated with `\left` and `\right` allow you to make specifiers of arbitrary size. These symbols can, if necessary, grow larger than symbols generated by `\Biggl`.

`\left` and `\right` must come in pairs. If you do *not* want to use a specific delimiter write "." for what is called the *null delimiter*. It typesets an empty box of width `\nulldelimiterspace` (default value 1.2pt).

While usually the delimiters of `\left` and `\right` match (for instance, opening and closing parenthesis, `\left(` and `\right)`), this is by no means required. You can have `\left)` matching a `(`, and you can have two totally unrelated symbol in a pair of `\left` and `\right`. See the following example for what you can have.

13.6.3 An Example using `\left` and `\right`

The following examples show that indeed the `\left` and the `\right` instructions are very powerful, because they provide for delimiters that can grow arbitrarily large.

The following input was used for the example:

```
$$
\left\lbrace
    15 +
    {
        \left\Downarrow
            \sqrt{a-1} \over \sqrt{b+1}
        \right]^2
    \over
        \left\lfloor
            \left( x \over y \right.
        \over
            \left( a \over b \right)
```

```
        \right\rfloor
    }
\right[^{13}
$$
```

This input generates the following output:

$$\left\{ 15 + \frac{\Big\Downarrow \left[\frac{\sqrt{a-1}}{\sqrt{b+1}} \right]^2}{\left\lfloor \frac{\left(\frac{x}{y}\right.}{\left(\frac{a}{b}\right)} \right\rfloor} \right.\left[\vphantom{\frac{\frac{a}{b}}{\frac{a}{b}}}\right.^{13}$$

13.6.4 Why and When is Explicit Size Specification Needed?

Now we address the question why one needs the capability to specify the size of a delimiter explicitly when TEX is obviously able to compute the size of delimiters using `\left` and `\right`. There are two reasons:

1. Sometimes TEX doesn't find the right size (its delimiters might be too small or too large).
2. There are cases where huge formulas have to be broken up over two or more separate lines. Then the "`\left`... `\right`" construct cannot be applied, because matching `\left` and `\right` must all appear within the same formula.

Here is a brief example. In order to typeset the following formula

$$\begin{aligned} A = & \Biggl(\pi \Biggr. \\ & + \frac{a+b+c+d}{\frac{d}{e}} \\ & + \frac{\frac{x}{y}}{a+b} \Biggl. \Biggr)^2 \end{aligned}$$

you have to enter

```
$$
    \eqalign {A = & \Biggl( \pi \Biggr.\cr
                  & + {a+b+c+d\over {d\over e}}\cr
                  & + {{x\over y}\over a+b}\Biggl.\Biggr)^2\cr
             }
$$
```

If you use `\left` and `\right`, as shown in the following source code,

```
$$
    \eqalign {A = & \left( \pi\right.\cr
                  & + {a+b+c+d\over {d\over e}}\cr
```

```
                 & + \left.{{x\over y}\over a+b} \right)^2\cr
           }
$$
```

then the output is incorrect (the source code is *legal*, of course), because the opening parenthesis and the closing parenthesis are of different size. On the other hand, the input

```
$$
     \eqalign {A = & \left( \pi\cr
                 & + {a+b+c+d\over {d\over e}}\cr
                 & + {{x\over y}\over a+b} \right)^2\cr
           }
$$
```

is illegal, because `\left` and `\right` must close on the same level (the same line).

For more on `\eqalign` see 14.10.1, p. 220. The other way to typeset such a formula using `\left` and `\right` is to insert a phantom (a strut-like construct).

13.6.5 Fine Points in the Computation of Delimiter Sizes, `\delimitershortfall`, `\delimiterfactor`

When TEX computes the actual size of a delimiter, then it first computes the size of the formula around which delimiters are to be inserted. Next is the size computation of the delimiter itself. Assume that the formula around which delimiters are to be inserted is h high and d deep. Compute $x = \max(h, d)$. `\delimiterfactor` f is a magnification times 1000. The first approximation of the length of the delimiter becomes $2 * x * (f/1000)$. The second approximation is $2 * y - \delta$, where δ is the value of `\delimitershortfall`, a dimension. The larger of the two approximations determines the vertical length of the delimiter.

13.7 Horizontal Spacing in Math Modes

Usually TEX's horizontal spacing in equations is very good. However, there are instances, where you need to intervene manually.

The TEXbook (on page 170) contains a table which describes TEX's built-in spacing rules. Eight different classes of atoms are relevant. Given a left and a right atom, the space inserted between the two atoms can be read from this table.

13.7.1 Changing Spacing by Hand

First here are the ways of changing TEX's horizontal spacing in math equations manually.

1. You can use `\hskip` inside mathematical equations. From that follows, of course, that you can use any macros which are built on `\hskip` such as `\quad` and `\qquad`. `\hskip` is discussed in detail in 5.1, p. I-121. For instance, `$A B \quad C$` prints $AB \quad C$.
2. `\␣`, control space, also works in math mode. For instance, `$AB\ C$` prints $AB\ C$.
3. `\kern` (horizontal kern) can be used too. See 5.4.8, p. I-140, for a discussion of kerning.
4. There is a special math glue which is generated using `\mskip`. The main difference between `\mskip` and `\hskip` is that `\mskip` can *only* refer to the dimension unit "mu." The unit *mu* is defined as follows: there are 18 mu to 1 em, where the unit em is taken from family 2, the math symbols family.

   The following macros based on math glue are defined in the plain format. First three math glue registers need to be initialized as follows:

   ```
   \thinmuskip = 3mu
   \medmuskip = 4mu plus 2mu minus 4mu
   \thickmuskip = 5mu plus 5mu
   ```

   Here is the list of macros based on `\mskip` glue.

   - `\,` is defined as `\def\,{\mskip\thinmuskip}`.
   - `\>` is defined as `\def\>{\mskip\medmuskip}`.
   - `\;` is defined as `\def\;{\mskip\thickmuskip}`.
   - `\!` is defined as `\def\!{\mskip-\thinmuskip}`.

5. Finally there is `\mkern`, math kern, which is very similar to horizontal kern (`\kern`) except that unit mu must be used (similar to the way `\mglue` is dealt with).

Here are some of the instances in which you need to correct the mathematical spacing.

13.7.2 Instances Where Corrections in the Horizontal Spacing May Be Appropriate

We will now list a few examples where a correction in the spacing is appropriate. Glitches like the following can usually only be fixed after the first output. If they presented fundamental problems with TEX one could program these rules into TEX and this way would not have to worry about it.

1. Insert some extra thin space before the dx, dy, d, or whatever of an integral. For instance, `$\int_0^\infty f(x)\,dx$` results in $\int_0^\infty f(x)\,dx$.
2. When *physical units* appear in a formula, they should be set in roman type and separated from the formula by a thin space. Type `$55\rm\, mi/hr$` to get $55\,\rm mi/hr$.
3. Insert a thin space after the exclamation point of a *factorial*. For example, enter `$(2n)!\,(2m)!$` to get $(2n)!\,(2m)!$.
4. Sometimes braces which denote sets need some tuning of their spacing—see 14.6, p. 207, for details.

13.8 Inline Mathmode

There are two items which apply to inline mathmode only which are the subject of this Section: one is additional glue inserted before and after inline equations, and the other is the issue of line breaks within inline equations.

13.8.1 Horizontal Glue Around Inline Equations, `\mathsurround`, `\m@th`

TEX, in order to offset an inline equation slightly, inserts horizontal glue in the amount of `\mathsurround` before and after each inline equation. The default of this glue parameter is 1.6 pt.

The macro `\m@th` sets the value in this register to zero.

13.8.2 Linebreaks in Inline Equations, `\relpenalty`, `\binoppenalty`

Because TEX analyzes a mathematical formula's structure before it does its typesetting, it prevents line breaks in the middle of an expression like $f(a,b)$, as it should. TEX can only work properly if you input equations properly, of course. If you entered `$f(a$, $b)$` instead of `$f(a,b)$`, then TEX will, of course, break the formula in the middle, if necessary. In general the rule is: keep things together which belong together logically. For instance, in "$f(a,b)$" the $f(a,b)$ is one formula and therefore should be entered in such a way.

There is the possibility of line breaks in the middle of a line, as the following formula shows: $f(a,b) = f_1(a,y) + f_2(1,2) + g'_{2,3} + h(f(1,2)) + l(k,0) + q(1) + l(y)+z(t)$. The question is where will TEX allow line breaks? The answer is quite simple: around binary and relational operators which occur on the outermost level. In those cases the parameters `\relpenalty` (default value of 500 in the plain format) and `\binoppenalty` (default of 700) determine horizontal penalties which determine the line breaks. An explicit penalty specification (like `\penalty 2000`) can be also used. See 13.8.2 on the previous page for details on how horizontal penalties influence line breaks.

From the preceding discussion it follows, that in general punctuation should be outside an inline equation, unless it belongs to the equation logically. For instance, in the text "$f(a, b)$, $g(x, y)$ are two nice example formulas" the correct input is ``` ``$f(a,b)$, $g(x,y)$ ...'' ```, switching back to text mode for the insertion of a comma between the two formulas.

13.9 Summary

In this chapter we learned:

- The typesetting of mathematical equations in TEX is fun, relatively easy and delivers pleasant output.
- TEX has two different math modes: an inline math mode (the input to such a math equation is surrounded by simple dollar signs) and display math mode (the input to such an equation is surrounded by double dollar signs). Inline math equations appear within a paragraph. Displayed equations are printed centered, on a line by themselves.
- Superscripts and subscripts are printed in the proper size automatically by TEX. They are entered using ^ and _ respectively. A superscript or subscript that is longer than one token must be enclosed inside curly braces. Simultaneous superscripts and subscripts are allowed and they can be specified in any order.
- Empty lines are not allowed inside mathematical equations.
- Ordinary math variables in equations are set in italics. For certain operations, such as lim, roman font is required and TEX defines control sequences such as `\lim` to achieve this.
- Different alphabets are available in math mode, such as Greek letters and a calligraphic alphabet.
- TEX knows many special control sequences specific to math mode. For instance, `\root` generates a root sign, `\over` is used for the generation of fractions, `\choice` for the generation of binomial coefficients, and so forth.
- TEX classifies symbols and operators in various classes. Tables which list the available operators and other special symbols (such as arrows) were presented.
- It is possible to overline and underline symbols.
- Text in mathematical equations require the text to be either inserted into an hbox or explicit spaces are necessary, because by default TEX ignores spaces in math mode.
- TEX has a separate set of math accents (separate from text accents). TEX's math mode has a special set of wide accents.

- TEX has a number of large operators such as summation and integration. Superscript and subscript syntax is used to enter the limits which apply to a large operator, such as to a sum or an integral. The placement of these limits can be controlled with `\limit` and `\nolimit`.
- Delimiters in math equations such as parentheses size automatically if `\left` and `\right` are used. They can be scaled manually using control sequences such as `\bigl`, `\Bigl` and so forth.
- The horizontal spacing in math equations is usually very good, but needs to be corrected occasionally using macros such as `\;` and `\>`.

14 More on Math

Now it's time to get down to some nitty gritty details in the processing of mathematical equations.

14.1 Converting Math Formulas Into Printed Output

This and the following four Sections describe the internals of TeX when typesetting mathematical formulas. We will investigate so-called *math lists* which are built by TeX. They are used as a basis to convert a formula into boxes and ultimately printed output. We will also discuss the concept of *math code* and *font families*.

14.1.1 A Math List Example, `\showlists`

For this Section we will use the following formula for an example:

$$\sum \frac{x_3^2}{a+b}$$

Even though we do not understand at this point what a math list is, let us see how we can generate one. Here is the source code to generate such a math list (to write it to the log file).

• **`ex-showlists.tip`** •

The `\nonstopmode` prevents TeX from halting after the `\showlists` output. The two `\showbox...` parameters are set to high values to allow for the full display of the math list generated below. See 4.5.14, p. I-105, for details.

```
\nonstopmode
\showboxbreadth = 10000
```

```
\showboxdepth    = 10000
$
    \sum
    {
        x^{2}_{3}
    \over
        a+b
    }
    \showlists
$
\bye
```

• End of **ex-showlists.tip** •

The log file (slightly shortened) generated by the preceding code reads as follows.

• **ex-showlists.log** •

```
1  This is TeX, C Version 3.14 (...)
2  **&/usr/local/tex/lib/fmt/plain ex-showlists.tip
3  (ex-showlists.tip
4
5  ### math mode entered at line 4
6  \mathop
7  .\fam3 P
8  \mathord
9  .\fraction, thickness = default
10 .\\mathord
11 .\.\fam1 x
12 .\^\fam0 2
13 .\_\fam0 3
14 ./\mathord
15 ./.\fam1 a
16 ./\mathbin
17 ./.\fam0 +
18 ./\mathord
19 ./.\fam1 b
20
21 [1] )
22 Output written on ex-showlists.dvi (1 page, 444 bytes).
```

The preceding log file is to be interpreted as follows:

- `\mathop` (line 6) specifies one of the classes that formulas and characters are classified in. In our example, this pertains to the summation sign.
- A dot indicates the nucleus of an atom (for now think of an atom as just one of the list elements). The next line (line 7), for instance, contains an atom. The font being used is from font family 3, and the character `P` is printed. This character `P` is not to be interpreted literally here. The ASCII code of the `P` is `"50`, and the summation sign happens to be at that position in the fonts of family 3. See the definition of `\sum`, item 4, p. 194, which assigns class 1 (that is `\mathord`), family 3 and character code `"50` to `\sum`.

- A ^ is used for the superscript of an atom.
- A _ is used for the subscript of an atom.
- The superscript and subscript specification of an atom, if empty, is omitted.
- Other things you can identify in this list are **\mathord** (class 0, ordinary) and **\mathbin** (binary operations) type of elements.

14.1.2 The Building of a Math List

Using the preceding example formula we will now discuss how a math list is built, and what its purpose is.

1. Similar to the way TEX builds horizontal and vertical lists, TEX builds a *math list*[1]. This math list building starts when the opening math symbol ($ or $$) is discovered. It ends when the closing math symbol is encountered.
   A math list is a *collection* of the following items:
   - *Atoms.* Atoms will be discussed shortly. A full discussion of atoms can be found in 14.2 on the next page.
   - A *generalized fraction* resulting from **\above**, **\over**, etc.
   - A *boundary* from a **\left** or **\right**.
   - A *four-way choice* resulting from a **\mathchoice**.
   - A *style change* resulting from a user-inserted style change using a command such as **\displaystyle**, **\scriptstyle**, etc. Note: this refers to user-inserted style changes only. User styles changes are the exception and style change computations are usually done by TEX automatically.
   - *Other elements* such as horizontal material (rule, discretionary, penalty, whatsit), vertical material (**\mark**, **\vadjust**, **\insert**), glue (**\hskip**, **\mskip**, **\nonscript**, and kern (**\kern**, **\mkern**).
     These items listed under *other elements* become relevant, once the math list is converted into a horizontal list. It is during this process that the actual typesetting of a math formula happens. Our concern here is primarily the *construction* of the math list; "other elements" are of minor interest here.

2. Atoms are among the most important items of math lists.
   - Atoms consist of three *fields* the *nucleus*, *subscript* and *superscript*.
   - The nucleus of an atom contains a *type*. For instance, in our formula x, ^{2}, $_3$, a and b would all be of type *Ord* (for ordinary). We also have an atom representing the summation sign, etc.

[1] For the computer scientist: a math list is nothing else than TEX's version of a syntax tree.

- In our example the superscript and subscript fields of the atoms representing ^{2}, ^{3}, a, b, and $+$ are empty. The superscript and subscript fields of x point to the atoms representing ^{2} and $_3$.

3. When the closing math delimiter is encountered, the math list is traversed to convert the math list into boxes. More precisely:
   (a) *Style information* is appended to the math list atoms. TEX has eight different styles two of which being T (text style) and S (script style). When typesetting an inline math equation, TEX starts with style T (that style is attached to the atom representing the fraction). This leads to style S being assigned to the atoms representing the numerator and denominator. This, in turn, leads to the scriptscript styles to be attached to ^{2} and $_3$, and so forth.
   (b) In addition to style information a *family number* is assigned to every symbol or character which is printed later. A family is a collection of three fonts. The three fonts are normally identical except for their sizes. For instance, in x^{y^z}, the three characters belong to the same family. The current style is used to select one of the three subfonts associated with a family (in this example each subfont, sizes 10 pt, 7 pt and 5 pt) is selected exactly once.
   (c) The style decision and family number determination s are the basis for the actual font used to print a symbol.
   (d) Finally the math list is converted into boxes. This process is described in detail in Appendix G of the TEXbook.

14.2 A Closer Look at Atoms

Recall that an atom consists of three fields, the *nucleus, subscript* and *superscript*. There are thirteen different atom types which are the following:

- *Ord* (ordinary atom, as, for instance, x, 2).
- *Op* (large operator atom, as, for instance, $\sum$).
- *Bin* (binary operator atom, as, for instance, $+$, $-$, $/$).
- *Rel* (relational operator atom, as, for instance, $>$)
- *Open* (opening atom, like, [).
- *Close* (closing atom, like]).
- *Punct* (punctuation atom, like ,).
- *Inner* (inner atom like an atom representing a fraction such as $x/2$).
- *Over* (overline atom as, for instance, in $\overline{x}$).
- *Under* (underline atom, as, for instance, in $\underline{x}$).
- *Acc* (accented atom, as, for instance, in $\hat{x}$).
- *Rad* (radical atom, as, for instance, in $\sqrt{x}$).
- *Vcent* (vbox to be centered, generated by `\vcenter`).

14.3 Assigning Class, Family and Character Code to a Character

Every character has a class, family and character code in the typesetting of mathematical equations. These assignments (also the assignments of classes to subformulas) is discussed in this Section.

14.3.1 Putting Together a Mathcode

A mathcode is defined for each character. How this math code is assigned with be discussed shortly. The subject here is the structure and the meaning of the math code (similar things apply to math symbols, see below):

1. Every character has the following information associated with in in math mode:

   (a) A class (a number in the range 0...7). The following classes are known in TEX.

Class	`\math...`	*Meaning*	*Example*
0	`\mathord`	Ordinary	`/`
1	`\mathop`	Large operator	`\sum`
2	`\mathbin`	Binary relation	`+`
3	`\mathrel`	Relation	`=`
4	`\mathopen`	Opening	`[`
5	`\mathclose`	Closing	`]`
6	`\mathpunct`	Punctuation	`,`
7		Variable family	`x`

   The second column containing `\math...` type of primitives, will be explained shortly.

   Class 7 is special in the way the family of the associated symbol is dealt with. See 14.5.5, p. 204, for details.

   (b) A family (a number in the range 0...15). See below for details.

   (c) A character code (a number on the range 0...127) in the selected family.

These three numbers are encoded into *one* hexadecimal number, four hex digits long, where

- The first digit represents the *class*.
- The second digit represents the *family*.
- The third and the fourth digits represent the *character code*.

That is class number, family number and character code are combined into a *mathcode* as follows: the mathcode of class x, family y and character code zz is the four digit long hexadecimal number $xyzz$.

2. A *subformula* is assigned a class only (no family and no character code is assigned). Usually TeX does these class assignments for you, but by using the instructions of the second column of the preceding class table you can force your own classes upon a subformula.

   For instance, `\mathbin{:}` would treat the colon as a binary operator. The colon does not need to be included in curly braces. If the formula, to which the classification was to be applied, is longer than one token, the curly braces are necessary, of course. The same rules as to undelimited arguments in macros apply (see 21.8.4, p. III-183).

   The various primitives listed in the preceding table can be used to set the type of a subformula to a specific value.

14.3.2 Assigning Mathcodes to Characters and Control Sequences, `\mathcode`, `\mathchar`, `\mathchardef`

The next questions that arise are the following:

1. How can one generate a math symbol with a specific class, family and character code in a math formula? The answer is very simple: use `\mathchar`. For instance, `$a \mathchar "313C b$` prints $a < b$, because class 3, family 1, character code `"3C` selects the less than sign, the way the plain format, and the Computer Modern fonts used with it, are set up.
2. How we can simplify things? Writing `\mathchar` instructions all the time would be awkward. First of all, we could use a macro and define `\lt` (for less than) as follows: `\def\lt{\matchar "313C }`.
3. What we really would like to do is *assign* the *math code* `"313C` to the `<` character. This can be done easily in TeX using `\mathcode` (note: not `\mathchar`).

   This primitive has the following format: `\mathcode` ⟨character code⟩ `=` ⟨mathcode⟩, and the specified mathcode is assigned to the character with the specified ⟨character code⟩. Here is an example (remember `` `< `` generates the character code of `<`):

```
\mathcode `< = "313C
```

   After this instruction was issued (usually as part of the format being used) you can use the character `<` directly, and achieve the desired effect. Thus writing `$a<b$` now prints $a < b$.

   The special mathcode value `"8000` makes a character behave like an active character; see 18.2.9, p. III-23, for details.
4. Given a control sequence such as `\sum`, how do we get TeX to generate $\sum$? Again this is very simple. The first possibility is to use `\mathchar`, and to simply define a macro.

For instance, to generate a summation sign, we need class 1, family 3, character code "50. We therefore could define macro \sum as follows:

```
\def\sum{\mathchar "1350 }
```

There is a shorter form though to do the same. \mathchardef can be used to assign a mathcode to a control sequence. Thus the definition of \sum in the plain format really reads \mathchardef\sum = "1350. \mathchardef and \mathchar relate the same way \chardef and \char do.

There is an eighth classification called \mathinner. It is normally used for fractions and \left...\right constructs. It adds some additional space to the left and right of the subformula, under certain circumstances.

14.4 Style Selection in Math Modes

After TEX has built the complete math list, TEX traverses this list assigning styles to the various list elements. The purpose of these styles is to facilitate determining the proper sizes of elements later. From the style of a symbol one of the three fonts of the symbol's family is selected. Note though while the font selection (within the three fonts of a family) and the style selection are related, the style selection is made first. Then the proper font is select based on style and family.

14.4.1 Styles

TEX knows about *eight* styles. These styles are D, T, S, SS and D', T', S' and SS'. Four of these styles are explained in the following table. The only difference between a style X and its cramped counterpart X' is that in the cramped styles superscripts are raised a little less than in the uncramped style.

The following four instructions of TEX allow the user to manually overwrite automatic style selection:

- \displaystyle,
- \textstyle,
- \scriptstyle,
- \scriptscriptstyle.

There are no instructions to change to one of the cramped styles. See 14.4.4,

p. 197, for when you need to select styles manually.

Style code	*Name*	*Command*
D	display style	`\displaystyle`
T	text style	`\textstyle`
S	script style	`\scriptstyle`
SS	scriptscript style	`\scriptscriptstyle`

Display style and text style are the respective styles TEX starts out with in display math mode and text mode. TEX selects script style for the superscript or subscript of something in display style or text style. The exact rules are the subject of 14.4.3 on this page.

14.4.2 Styles and Sizes

TEX has three different font sizes for the typesetting of math formulas. These sizes are called *text size*, *script size* and *script script size*. Each family contains three fonts, and each font in a font family represents one of those sizes. Assume that you know the *style* of a formula, then the following table allows you to determine the *size* which will be used to typeset the formula.

Style of formula	Size which will be used
D, D', T, T'	text size
S, S'	script size
SS, SS'	scriptscript size

14.4.3 Rules for the Selection of Styles

Next let us discuss how the *styles* are selected once the math list has been built.

- *Rule 1.* The style selected at the beginning of an inline math mode equation is T. For a display math mode equation TEX starts using style D.
- *Rule 2.* This rule defines the styles selected for *superscripts* and *subscripts*:

Style of *formula*	Style of *superscript*	Style of *subscript*
D, T	S	S'
D', T'	S'	S'
S, SS	SS	SS'
S', SS'	SS'	SS'

- *Rule 3.* This rule defines the styles of the *numerator* and *denominator* of a fraction, given the style of a subformula:

Style of *formula*	Style of *numerator*	Style of *denominator*
D	T	T'
D'	T'	T'
T	S	S'
T'	S'	S'
S, SS	SS	SS'
S', SS'	SS'	SS'

- *Rule 4.* `\underline` does not change the style.
- *Rule 5.* Math accents, `\sqrt`, and `\overline` change their uncramped styles to their cramped counterparts.

In addition to the preceding rules which describe the way TEX determines styles automatically, you can use the instructions of 14.4.1, p. 195, to put your own styles in place.

14.4.4 Manual Style Selections

There are cases where it is necessary for the user to select the proper styles. This causes special style change elements to be inserted into the math list. Here is an example where manual style selection (and the insertion of struts) is necessary.

For instance, first you might enter

```
$$
x_0 + {1\over x_1+
        { 1\over x_2+
          { 1\over x_3+
            { 1\over x_4+
              { 1\over x_5}}}}}
$$
```

and be surprised about the not-so-great-looking result:

$$x_0 + {1\over x_1+{1\over x_2+{1\over x_3+{1\over x_4+{1\over x_5}}}}}$$

Certainly,

$$x_0 + \cfrac{1}{x_1 + \cfrac{1}{x_2 + \cfrac{1}{x_3 + \cfrac{1}{x_4 + \cfrac{1}{x_5}}}}}$$

looks much better which was generated by the following source code:

```
$$
x_0 + {1\over\displaystyle x_1+
        {\strut 1\over\displaystyle x_2+
          {\strut 1\over\displaystyle x_3+
            {\strut 1\over\displaystyle x_4+
              {\strut 1\over\displaystyle x_5}}}}}
$$
```

The inserted strut (`\strut`) in the preceding formula ensures constant and consistent vertical spacing (one of the few moments where it is necessary to intervene by hand).

The `\displaystyle` instruction was used frequently in this and the next chapter, because many examples of *displayed* equations are parts of tables. One column contains the source code, another column contains the corresponding output (in display math mode). Of course, the output is made part of a column, and *not* centered over the whole line. To generate the output in this output column the following trick is used: use inline math mode, but force display style as initial style. `$\displaystyle F$` does just that to formula F.

14.4.5 Making Up Your Own Symbols, `\mathchoice`, `\mathpalette`

There are cases where you would like to define your own symbol, and where this symbol is supposed to change (usually in size) depending on what the current style is. For that purpose primitive `\mathchoice` can be used. It is applied as follows:

`\mathchoice` {⟨math⟩}{⟨math⟩}{⟨math⟩}{⟨math⟩}

The first subformula is selected for styles D and D', the second for T and T', the third for S and S', and the fourth for SS and SS'.

For instance, in the following example `\x` will print either a 0, 1, 2, or 3 depending on the current style (usually, of course, the same symbol is printed, just at different sizes):

```
$$
     \def\x{\mathchoice{1}{2}{3}{4}}
     \x + {\x \over 20} + \x^{\x} + \x ^{\x ^{\x}}
$$
```

The preceding source code generates the following output:

$$1+\frac{2}{20}+1^3+1^{3^4}$$

Often it is more convenient to use the plain format macro `\mathpalette`. This macro has two parameters:

- #1. This argument must be a control sequence which itself has two arguments, the first one being a style selection (`\displaystyle`, etc). The second argument of #1 is the second argument of `\mathchoice`.
- #2. See the explanation of #1.

```
\def\mathpalette #1#2{%
    \mathchoice
        {#1\displaystyle{#2}}%
        {#1\textstyle{#2}}%
        {#1\scriptstyle{#2}}%
        {#1\scriptscriptstyle{#2}}%
}
```

For example, `\mathpalette {\Q}{123}` expands to

```
\mathchoice{\Q\displaystyle{123}}
           {\Q\textstyle{123}}
           {\Q\scriptstyle{123}}
           {\Q\scriptscriptstyle{123}}
```

14.5 Fonts and Font Families in Math Mode

We will now discuss the font management in math formula typesetting. All characters and symbols in math mode belong to one of up to 16 *families*. Families are identified by a number in the range $0 \ldots 15$, which we call the *family index*.

14.5.1 Loading Math Fonts in the Plain Format

We will first discuss which special fonts are loaded for the typesetting of mathematical formulas in the plain formats. Other formats may load additional fonts, but what follows applies usually to all format. Note that *loading* the following fonts is accomplished by using `\font` (see 15.2.9, p. 238, for details).

1. *Computer Modern Math Italic* (`cmmi...`). This font is a math italic font. Its layout can be found in Figure 14.1 on the next page. This font is accessible in the math mode of the plain format through family 1 (`\mit`).

```
\font\teni = cmmi10
\font\preloaded = cmmi9
\font\preloaded = cmmi8
\font\seveni = cmmi7
\font\preloaded = cmmi6
\font\fivei = cmmi5
```

	´0	´1	´2	´3	´4	´5	´6	´7	
´00x	Γ	Δ	Θ	Λ	Ξ	Π	Σ	Υ	"0x
´01x	Φ	Ψ	Ω	ff	fi	fl	ffi	ffl	
´02x	ı	ȷ	`	´	ˇ	˘	¯	˚	"1x
´03x	¸	ß	æ	œ	ø	Æ	Œ	Ø	
´04x	-	!	"	#	$	%	&	'	"2x
´05x	(	)	*	+	,	-	.	/	
´06x	0	1	2	3	4	5	6	7	"3x
´07x	8	9	:	;	¡	=	¿	?	
´10x	@	A	B	C	D	E	F	G	"4x
´11x	H	I	J	K	L	M	N	O	
´12x	P	Q	R	S	T	U	V	W	"5x
´13x	X	Y	Z	[	"	]	^	˙	
´14x	'	a	b	c	d	e	f	g	"6x
´15x	h	i	j	k	l	m	n	o	
´16x	p	q	r	s	t	u	v	w	"7x
´17x	x	y	z	–	—	˝	~	¨	
	"8	"9	"A	"B	"C	"D	"E	"F	

Figure 14.1. Math italic font layout.

2. *Computer Modern Math Symbol* (cmsy...). This font contains most of the math symbols. It is accessible through font family 2. Its layout can be found in Figure 14.2 on the next page.

```
\font\tensy = cmsy10
\font\preloaded = cmsy9
\font\preloaded = cmsy8
\font\sevensy = cmsy7
\font\preloaded = cmsy6
\font\fivesy = cmsy5
```

3. *Computer Modern Math Extension* (cmex10). This font contains additional math symbols, in particular those characters used to print large operators, root signs, etc. This font is accessible through font family 2 in the plain format. It is only loaded in a 10 pt size, because it contains the large and the small version of the relevant symbols such as the summation sign and so forth. Its layout can be found in Figure 14.3, p. 202.

```
\font\tenex = cmex10
```

	'0	*'1*	*'2*	*'3*	*'4*	*'5*	*'6*	*'7*	
'00x	$-$	$\cdot$	$\times$	$\ast$	$\div$	$\diamond$	$\pm$	$\mp$	"0x
'01x	$\oplus$	$\ominus$	$\otimes$	$\oslash$	$\odot$	$\bigcirc$	$\circ$	$\bullet$	
'02x	$\asymp$	$\equiv$	$\subseteq$	$\supseteq$	$\leq$	$\geq$	$\preceq$	$\succeq$	"1x
'03x	$\sim$	$\approx$	$\subset$	$\supset$	$\ll$	$\gg$	$\prec$	$\succ$	
'04x	$\leftarrow$	$\rightarrow$	$\uparrow$	$\downarrow$	$\leftrightarrow$	$\nearrow$	$\searrow$	$\simeq$	"2x
'05x	$\Leftarrow$	$\Rightarrow$	$\Uparrow$	$\Downarrow$	$\Leftrightarrow$	$\nwarrow$	$\swarrow$	$\propto$	
'06x	$\prime$	∞	$\in$	$\ni$	$\triangle$	$\triangledown$	$/$	$\mapstochar$	"3x
'07x	$\forall$	$\exists$	$\neg$	$\emptyset$	$\Re$	$\Im$	$\top$	$\perp$	
'10x	$\aleph$	$\mathcal{A}$	$\mathcal{B}$	$\mathcal{C}$	$\mathcal{D}$	$\mathcal{E}$	$\mathcal{F}$	$\mathcal{G}$	"4x
'11x	$\mathcal{H}$	$\mathcal{I}$	$\mathcal{J}$	$\mathcal{K}$	$\mathcal{L}$	$\mathcal{M}$	$\mathcal{N}$	$\mathcal{O}$	
'12x	$\mathcal{P}$	$\mathcal{Q}$	$\mathcal{R}$	$\mathcal{S}$	$\mathcal{T}$	$\mathcal{U}$	$\mathcal{V}$	$\mathcal{W}$	"5x
'13x	$\mathcal{X}$	$\mathcal{Y}$	$\mathcal{Z}$	$\cup$	$\cap$	$\uplus$	$\wedge$	$\vee$	
'14x	$\vdash$	$\dashv$	$\lfloor$	$\rfloor$	$\lceil$	$\rceil$	$\{$	$\}$	"6x
'15x	$\langle$	$\rangle$	$\vert$	$\Vert$	$\updownarrow$	$\Updownarrow$	$\backslash$	$\wr$	
'16x	$\surd$	$\amalg$	∇	$\smallint$	$\sqcup$	$\sqcap$	$\sqsubseteq$	$\sqsupseteq$	"7x
'17x	$\S$	$\dagger$	$\ddagger$	$\P$	$\clubsuit$	$\diamondsuit$	$\heartsuit$	$\spadesuit$	
	"8	"9	"A	"B	"C	"D	"E	"F	

Figure 14.2. Math symbol font layout.

14.5.2 The Fonts of a Family, `\textfont`, `\scriptfont`, `\scriptscriptfont`

Font families group together *three fonts*. These three fonts are identified as follows:

1. **`\textfont`** identifies a font for *text size.*
2. **`\scriptfont`** identifies a font for *script size.*
3. **`\scriptscriptfont`** identifies a font for *scriptscript size.*

See also the table in 14.4.2, p. 196. Given a ⟨family number⟩ and a ⟨font identifier⟩ (acquired by a **`\font`** instruction to load a font) the three fonts of a family are assigned as follows:

> **`\textfont`** ⟨family number⟩ = ⟨font identifier⟩
> **`\scriptfont`** ⟨family number⟩ = ⟨font identifier⟩
> **`\scriptscriptfont`** ⟨family number⟩ = ⟨font identifier⟩

In case the **`\scriptfont`** or **`\scriptscriptfont`** of a family has *not* been defined, the **`\textfont`** of the family is used instead. It should be clear from the preceding

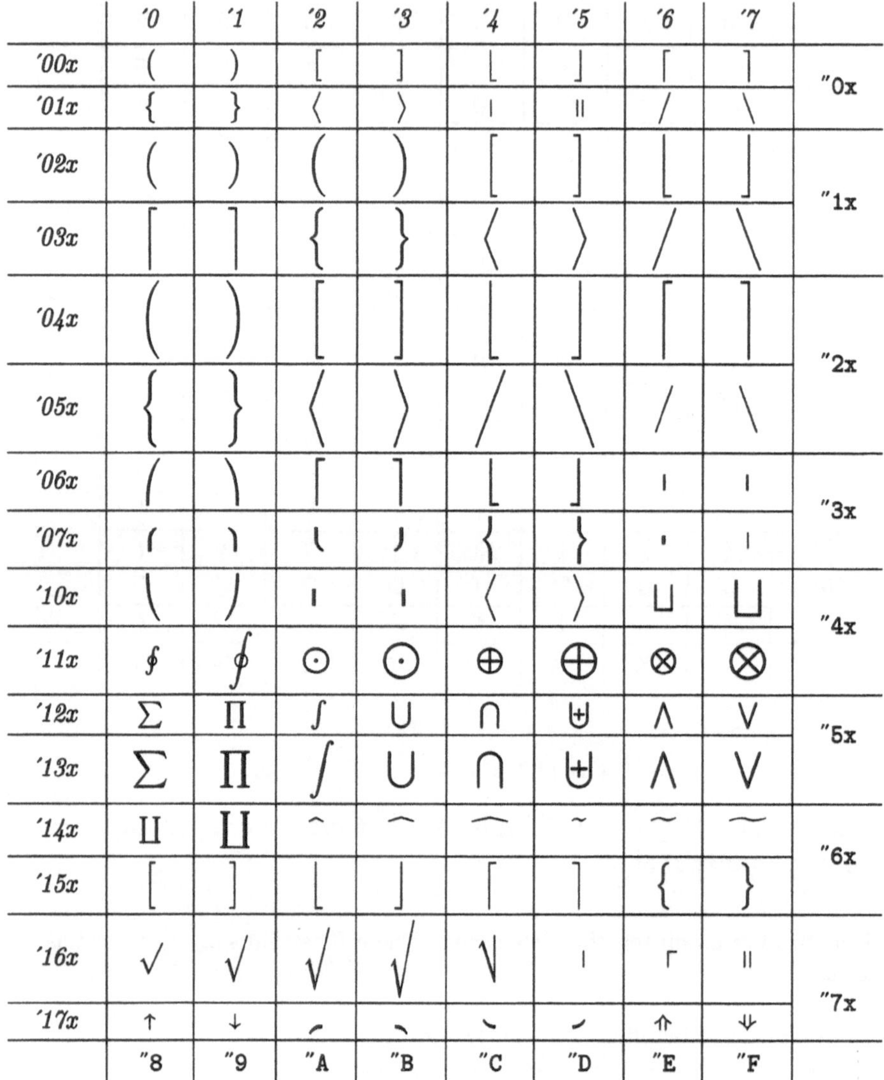

Figure 14.3. Math extension font layout.

discussion that in any reasonable setup, the three fonts of one family use the same font in different sizes.

Next we discuss the various families. Then we show an example of how loaded fonts are connected to the various families using the instructions of this Subsection.

14.5.3 The Meaning of the Various Font Families

In this Subsection we will briefly explain the meanings of the different font families. The font family numbers as defined below are *burned* into the TEX program. This means that there is no reasonable way for the user to assign different font families and still get reasonable results.

In addition to that the user *can* add his or her own fonts, by adding a new font family. This will be discussed later.

First here is the list of the fixed font families that the TEX program expects to be able to access for the typesetting of math formulas.

- *Family 0.* It is assumed that a regular roman text font is assigned to this family. This family is used for non-math variables (math variables are set in an italic font). For instance, the "lim" in "$\lim_{x>0}$" uses this font family.
- *Family 1.* It is assumed that a *math italic font* is assigned to this family. This family is used in particular for all variables in math formulas.
- *Family 2.* It is assumed that a *math symbol font* is assigned to this position.

  This family also defines the unit em used to define "mu," the basic unit of glue in math mode; see 13.7.1, item 4, p. 185.
- *Family 3.* It is assumed that a *math extension font* is assigned to this family. This family's purpose is to typeset large symbols.

The following font families are *not* fixed (in that they have some predetermined meaning), that is no specific font families are assumed by the TEX program to be available. We list here which font families are used by the plain format. From this it follows that font families 8–15 are unused in the plain format.

- *Family 4* contains *text italic.*
- *Family 5* contains *slanted roman.*
- *Family 6* contains **bold roman.**
- *Family 7* contains `typewriter type.`

14.5.4 Relating Sizes and Fonts, `\newfam`

In math mode, one does not have an active font, but an active *font family.* How TEX selects families will be discussed shortly. The preceding Subsection contains an overview of the available families.

Now let's say you decided that font family 0 is used for roman letters (as is explained in the preceding Subsection). To set up TEX in such a way, you need to do two things:

- Load the necessary fonts. You need to load three fonts for the three sizes which are used in TEX. The plain format sticks to a 10 pt, 7 pt, 5 pt scale. For that purpose use the `\font` instruction (see 15.2.9, p. 238):

```
\font\tenrm = cmr10
\font\sevenrm = cmr7
\font\fiverm = cmr5
```

- Tell TEX what the three fonts are for the particular font family using `\textfont`, `\scriptfont` and `\scriptscriptfont` as the following example shows:

```
\textfont 0 = \tenrm
\scriptfont 0 = \sevenrm
\scriptscriptfont 0 = \fiverm
```

With `\newfam` you can define a symbolic name for a family not yet used. For instance, `\newfam\myfam` would do that. Then you can write something like `\texfont\myfam = ...` and assign a font of your choice. The first font family assigned by a `\newfam` call in the plain format will be family 8, because families 0–7 are used by the plain format.

14.5.5 Integer Parameter `\fam`, Special Class 7

One question not answered in the preceding Subsection is, of course, how TEX *selects* the proper family. First of all you can force a specific family by writing, for instance, `\fam = 3`. But this is about the last thing you probably want to do. Before we discuss this, we need to discuss the special class 7.

A symbol in class 7 is treated in a special way, because this class allows the symbol to *change* the family according to the following rules. What happens depends on the current value of `\fam`.

1. If the value of `\fam` is -1, then classes 7 and 0 are equivalent, and the symbol is treated as if it belonged to class 0. Regardless of what family was assigned to the symbol, family 0 is used.
2. If the value of `\fam` is ≥ 0 (and < 15, that is a legal family value), then this integer parameter `\fam` specifies the family to be used.
3. Whenever math mode is started, TEX sets `\fam` to a value of -1.

14.5.6 The Math Family Definitions of the Plain Format

We will now present the math family definitions of the plain format. The font changing instructions (`\bf`, `\it`, and so forth) were introduced to you as ordinary font changes so far. In reality, in the plain format they are macros. Each of these macros consists of two parts:

- The first part involves a font family change.
- The second part involves a font change.

Observe that the first part (changing font families) is ignored by TEX when the instruction is used inside text, but is relevant inside math formulas. On the other hand, the second part of the instruction (changing fonts) is irrelevant for TEX inside math formulas, so that only the change of font families (first part) counts in math formulas.

Here are the math family definitions of the plain format.

- Family 0:

```
\textfont0 = \tenrm
\scriptfont0 = \sevenrm
\scriptscriptfont0 = \fiverm
\def\rm{\fam = 0 \tenrm}
```

- Family 1:

```
\textfont1 = \teni
\scriptfont1 = \seveni
\scriptscriptfont1 = \fivei
\def\mit{\fam = 1 }
\def\oldstyle{\fam = 1 \teni}
```

- Family 2:

```
\textfont2 = \tensy
\scriptfont2 = \sevensy
\scriptscriptfont2 = \fivesy
\def\cal{\fam = 2 }
```

- Family 3:

```
\textfont3 = \tenex
\scriptfont3 = \tenex
\scriptscriptfont3 = \tenex
```

- Family 4:

```
\newfam\itfam
\def\it{\fam = \itfam \tenit}
\textfont\itfam = \tenit
```

- Family 5:

```
\newfam\slfam
\def\sl{\fam = \slfam\tensl}
\textfont\slfam = \tensl
```

- Family 6:

```
\newfam\bffam
\def\bf{\fam = \bffam\tenbf}
\textfont\bffam = \tenbf
\scriptfont\bffam = \sevenbf
\scriptscriptfont\bffam = \fivebf
```

- Family 7:

```
\newfam\ttfam
```

```
    \def\tt{\fam = \ttfam\tentt}
    \textfont\ttfam = \tentt
```

The plain format obviously employs a 10 pt, 7 pt, 5 pt scaling.

14.5.7 Redefining Math Font Families

Here is an example where the various fonts of each family are made identical. This makes for pretty ugly looking formulas, because all characters now have the same size. But it also makes for a pretty good example. Assume, for instance, you wanted to define a setup of math fonts in such a way that formulas in footnotes are printed in the appropriate size. The example below shows that it is very easy and straightforward to redefine fonts under those circumstances.

The following macro uses 10 pt fonts for all fonts.

$\mathcal{P}'$ • ssmath.tip •

```
\def\SameSizeMath{
    \textfont0 = \tenrm
    \scriptfont0 = \tenrm
    \scriptscriptfont0 = \tenrm
    \textfont1 = \tenit
    \scriptfont1 = \tenit
    \scriptscriptfont1 = \tenit
    \textfont2 = \tensy
    \scriptfont2 = \tensy
    \scriptscriptfont2 = \tensy
    \textfont3 = \tenex
    \scriptfont3 = \tenex
    \scriptscriptfont3 = \tenex
    \textfont\itfam = \tenit
    \textfont\slfam = \tensl
    \textfont\bffam = \tenbf
    \scriptfont\bffam = \tenbf
    \scriptscriptfont\bffam = \tenbf
    \textfont\ttfam = \tentt
}
```

• End of ssmath.tip •

The example itself follows next.

• ex-ssmath.tip •

```
$$
    \SameSizeMath
    \sum_{i=1}^{200} x_{i+1}^{20^{30\over z-d}}
$$
```

• End of ex-ssmath.tip •

This input generates the following output:

$$\sum_{i=1}^{200} x_{i+1}^{20^{\frac{30}{z-d}}}$$

14.5.8 Adding Fonts to Math Mode

The proper way to add a font for the use in math mode is to define its own family using `\newfam` (see 14.5.4, p. 203). This is the cleanest approach because all sizing is done automatically then.

If you have only one or two characters from another font which you would like to use in a math equation, then you can use a short cut: enclose these characters in an hbox and switch to the appropriate font inside the hbox. Note that simply loading the font using `\font` and then switching to the font loaded will not work inside a math mode equation. By using an hbox TeX is put into restricted horizontal mode, and therefore regular font changing instructions will work.

14.5.9 `\defaultskewchar`

The math mode typesetting also defines a `\skewchar` for every font (default is the value of `\defaultskewchar`). For details see the TeXbook, page 442, which discusses in detail the conversion of formulas into boxes.

14.6 Braces

A brief discussion of braces, horizontal as well as vertical, is next.

14.6.1 Braces for Set Definitions

Because braces have a special meaning in TeX, you need to write `$\lbrace a, b, c \rbcase$` prints $\{a, b, c\}$. Also `$\{ a, b, c \}$` works, but I strongly recommend against using `\{` and `\}`, because it is very easy to mix up `\{` and `{`, or `\}` and `}`.

You should add some thin space (`\,`) in cases like $\{\, y \mid y > 5\,\}$, which was typeset by writing `$\lbrace\,y \mid y>5\,\rbrace$`.

14.6.2 Braces in Describing "Cases" in Math Formulas, `\cases`

Braces, which declare a choice in math, can be typeset as the following example shows (the input is nothing else than a special form of a two column table):

```
$$
f(x)=\cases{x^2+2,&           if $x > 0$\cr
            3,&               if $x = 0$\cr
            {a \over x}&      if $x < 0$\cr}
$$
```

This input generates the following output:

$$f(x) = \begin{cases} x^2+2, & \text{if } x > 0 \\ 3, & \text{if } x = 0 \\ \frac{a}{x} & \text{if } x < 0 \end{cases}$$

`\cases` typesets its own "{". There is no corresponding "}." The left part of the formula (that is the first column in table terminology) is printed automatically in math mode, whereas the second part is in regular text mode.

14.6.3 Braces Above and Below Formulas, `\overbrace` and `\underbrace`

There are some special braces which allow you to put a horizontal brace over a formula or below a formula in the following way. Here is an example:

Input		Output
`$$\overbrace{x+\cdots+x}^{k\rm\;times}$$`	prints	$\overbrace{x+\cdots+x}^{k\rm\;times}$
`$$\underbrace{x+y+z}_{>0}$$`	prints	$\underbrace{x+y+z}_{>0}$

14.7 Vertical Spacing, Phantoms, Struts in Math Formulas

Phantoms are used for fine manual control over the spacing of TEX. For instance, if you enter `$\sqrt{a} - \sqrt{b}$`, then the output reads $\sqrt{a}+\sqrt{b}$, with the top horizontal lines of the two square root signs on different horizontal levels. Entering `\mathstrut` with the two variables fixes the problem. The resulting output now reads $\sqrt{\mathstrut a}-\sqrt{\mathstrut b}$.

Here are the available instructions of the plain format:

`\phantom{`⟨formula⟩`}` generates an *empty box* of the *dimensions* of the given formula. In other words, it will do the typesetting as if you had just simply said ⟨formula⟩, but the formula will be invisible.

- `\vphantom{⟨formula⟩}` generates an *empty box* of the *height and depth of the formula* but of *width 0*.
- `\mathstrut` is defined as `\vphantom()`. The opening parenthesis is what determines the size of the generated phantom.
- `\smash{⟨formula⟩}` generates an *empty box* of *height and depth 0* but of the *width of the formula* which makes it orthogonal to `\phantom`. Actually that description more describes the effect than the implementation of this macro. Check the source code of the plain format for details.

14.8 How to Input Mathematical Equations

To compose a short inline math equation is rather straightforward. Larger display mathmode equation are more difficult. Structuring the TEX source is therefore very important.

The purpose of this Section is to look at the issue of how to structure the input to TEX's math mode more closely and to find ways of writing "readable" TEX source code.

14.8.1 Empty Lines Before and After Displayed Equations

An often neglected point is the following: *a displayed equation must neither be preceded nor followed by an empty line*, unless the user wants to indicate the start or end of a new paragraph. For instance, if the user writes the following input:

```
1 From
2 $$
3     \sum_{i=1}^n (x_i + y_i)  =  0
4 $$
5 follows that
6 $$
7     \sum_{i=1}^n x_i + \sum_{i=1}^n y_i =  0\ \ .
8 $$
9     This is consistent with our result in the previous chapter.
```

Then the three text lines (lines 1, 5, and 9) and the two equations in between them are part of one and the same paragraph, as they should be. If the user instead had written

```
From

$$
    \sum_{i=1}^n (x_i + y_i)  =  0
$$

```

```
follows that

$$
    \sum_{i=1}^n x_i + \sum_{i=1}^n y_i =  0\ \ .
$$

This is consistent with our result in the previous chapter.
```

then the texts “Follows that” and “This is consistent” all start new paragraphs. This inserts vertical glue `\parskip` glue before each of them. This leads to improper vertical spacing if its value is different from zero. It also leads to “follows that” indented by the paragraph indentation, which is definitely wrong.

14.8.2 Recommendations for How to Write Readable Math Source Code

The next question is how to input an equation in such a way that its source code can be read and corrected easily. This means structuring a formula in such a way that it closely resembles the final output. Here are a couple of suggestions (observe that spaces are irrelevant in math mode, and so the user can input an equation with white space anywhere where it helps the readability of the source code):

1. Write super- and subscripts close to the formula, without spaces.
2. Contrary to that leave spaces around binary operators like “+” and “–.”
3. For long fractions consider putting the numerator on one line, the `\over` on the next, followed by the denominator on the line after that.
4. Indent your equations in the source code.
5. Put `\left` and `\right` on different lines, indented by the same amount. This way you can find the matching pairs of those expressions easily.
6. In general try to input a formula as much as possible close to the way it comes out.
7. In case of aligned equations follow the rules for typing tables, see 39.7, p. IV-291. You may also want to look into 14.10.1, p. 220.

14.8.3 An Example

The above principles are now exemplified. Assume that you have to type the following equation:

$$\frac{\sum_{i=1}^{n}\left(x_i + \frac{y_i}{2}\right)}{\sum_{j=1}^{2n}\left(\frac{a_j}{2} + \frac{y_j}{2}\right)} = 2 + \frac{\sqrt{\frac{A+4}{2}}}{\pi}$$

Here is how you *should* enter this equation:

```
$$
        {
            \sum_{i=1}^n
                \left(
                     x_i + {y_i \over 2}
                \right)
        \over
            \sum_{j=1}^{2n}
                \left(
                     {a_j \over 2} + {y_j \over 2}
                \right)
        }
    =
        2 + {
                \sqrt{A+4 \over 2}
            \over
                \pi
            }
$$
```

And here is another way, giving you the same output, but how would you ever fix a problem when you entered the input in the following way:

```
$${\sum_{i=1}^n\left(x  _i+{y   _i\over2}
\right)        \over\sum_{j=1}^{2n}   \left(
{     a_j\over2}+{y_i\over
2}        \right)}=2+{              \sqrt{A+4 \over 2}
     \over\pi
} $$
```

14.8.4 Building a Formula Step-By-Step Using Macros

In the following we will show another method of how to approach a complicated formula. We take a formula, and typeset it in pieces by defining short macros, each typesetting such a piece. These small pieces are then combined into bigger pieces, again using macros. This is quite convenient, especially in the beginning, when you just start to learn the math mode of TEX.

As an example let us typeset the following formula:

$$\frac{F(x,y)}{2^n} = \sqrt{\frac{\displaystyle\sum_{j=1}^{n} \frac{\prod_{i=1}^{10} x_{i,j}}{\prod_{i=1}^{15} \frac{x_{i,j}}{y_{i,j}}}}{\sqrt{\sqrt{\int \int_a^b \frac{f(x)}{g(x)} dx}}}}$$

Define macros which typeset small pieces of the formula. Subsequently those macros are combined with other macros to form even bigger pieces. Here we go. Define the following macros:

```
\def\proda{\prod_{i=1}^{10}x_{i,j}}
\def\xiyi{{x_{i,j} \over y_{i,j}}}
\def\prodb{\prod_{i=1}^{15}\xiyi}
```

These macros generate the following output:

$$\texttt{\textbackslash proda} : \prod_{i=1}^{10} x_{i,j}, \quad \texttt{\textbackslash xiyi} : \frac{x_{i,j}}{y_{i,j}}, \quad \texttt{\textbackslash prodb} : \prod_{i=1}^{15} \frac{x_{i,j}}{y_{i,j}}$$

Next the following macros are defined:

```
\def\fracprod{{\proda \over \prodb}}
\def\asum{\sum_{i=1}^n \fracprod}
```

Here is the output generated by `\fracprod` and `\asum`.

$$\texttt{\textbackslash fracprod} : \frac{\prod_{i=1}^{10} x_{i,j}}{\prod_{i=1}^{15} \frac{x_{i,j}}{y_{i,j}}}, \quad \texttt{\textbackslash asum} : \sum_{i=1}^{n} \frac{\prod_{i=1}^{10} x_{i,j}}{\prod_{i=1}^{15} \frac{x_{i,j}}{y_{i,j}}}$$

Now we come to the denominator of the fraction under the square root.

```
\def\dint{\int\!\!\int}%
\def\dsqrt #1{\sqrt{\sqrt{#1}}}%
\def\ifrac{{f(x) \over g(x)}}%
\def\denom{\dsqrt{\dint_a^b \ifrac dx}}%
```

$$\texttt{\textbackslash ifrac} : \frac{f(x)}{g(x)}, \quad \texttt{\textbackslash denom} : \sqrt{\sqrt{\int\!\!\int_a^b \frac{f(x)}{g(x)} dx}}$$

Finally we can finish the right side of the formula

```
\def\rightside{\sqrt{{\asum \over \denom}}}
```

$$\texttt{\textbackslash rightside} : \sqrt{\frac{\sum_{i=1}^{n} \frac{\prod_{i=1}^{10} x_{i,j}}{\prod_{i=1}^{15} \frac{x_{i,j}}{y_{i,j}}}}{\sqrt{\sqrt{\int\!\!\int_a^b \frac{f(x)}{g(x)} dx}}}}$$

There is one final correction to be done and that is that it seems more desirable to print the summation over i in display style. This can be achieved the following way:

```
\let\asumold = \asum
\def\asum{\displaystyle\asumold}
```

Here is the final version of the right side of the formula:

$$\texttt{\textbackslash rightside} : \sqrt{\frac{\displaystyle\sum_{i=1}^{n} \frac{\prod_{i=1}^{10} x_{i,j}}{\prod_{i=1}^{15} \frac{x_{i,j}}{y_{i,j}}}}{\sqrt{\sqrt{\int\!\!\int_a^b \frac{f(x)}{g(x)} dx}}}}$$

The left side is done easily:

```
\def\leftside{{F(x,y) \over 2^{n}}}
```

$$\texttt{\textbackslash leftside} : \frac{F(x,y)}{2^n}$$

And now everything together prints the complete formula:

```
\leftside = \rightside
```

Here is the final output:

$$\frac{F(x,y)}{2^n} = \sqrt{\frac{\sum_{i=1}^{n} \frac{\prod_{i=1}^{10} x_{i,j}}{\prod_{i=1}^{15} \frac{x_{i,j}}{y_{i,j}}}}{\sqrt{\sqrt{\iint_a^b \frac{f(x)}{g(x)} dx}}}}$$

14.8.5 Stepwise Refinement

In the following we typeset the same formula as before, proceeding in a different way: we start with the outermost constructs and then continuously fill in pieces. It will be clear momentarily what we mean by that precisely.

Let us first start with typing the major structure of the formula which is a fraction and a square root:

```
    {
        X
    \over
        Y
    }
=
    \sqrt{
        frac
     }
```

The preceding code prints the following output:

$$\frac{X}{Y} = \sqrt{frac}$$

The next step is to fill in the left-hand side and to start filling in the right-hand side:

```
    {
        F(x,y)
    \over
        2^n
    }
=
```

```
     \sqrt{
         {
             SUM
         \over
             SQRT
         }
     }
```

Here is the output generated by the preceding source code:

$$\frac{F(x,y)}{2^n} = \sqrt{\frac{SUM}{SQRT}}$$

Now we continue working on the sum and the square roots in the denominator:

```
         {
             F(x,y)
         \over
             2^n
         }
     =
         \sqrt{
             {
                 \sum_{j=1}^n frac
             \over
                 \sqrt{\sqrt{integrals}}
             }
         }
```

$$\frac{F(x,y)}{2^n} = \sqrt{\frac{\sum_{j=1}^n frac}{\sqrt{\sqrt{integrals}}}}$$

More details are filled in.

```
         {
             F(x,y)
         \over
             2^n
         }
     =
         \sqrt{
             {
                 \sum_{j=1}^n {
                                 proda
                             \over
                                 prodb
                             }
             \over
                 \sqrt{
                     \sqrt{
                         \int\!\!\int_a^b frac dx
```

```
                    }
                }
        }
    }
```

$$\frac{F(x,y)}{2^n} = \sqrt{\frac{\sum_{j=1}^n \frac{proda}{prodb}}{\sqrt{\sqrt{\int\!\!\int_a^b fracdx}}}}$$

Even more here. Notice the carefully selected indentation.

```
    {
        F(x,y)
    \over
        2^n
    }
=
    \sqrt{
        {
            \sum_{j=1}^n {
                                \prod_{i=1}^{10} x_{i,j}
                            \over
                                \prod_{i=1}^{15} frac
                            }
        \over
            \sqrt{\sqrt{\int\!\int\!_a^b {
                                            f(x)
                                        \over
                                            g(x)
                                        }
                            dx
                        }
                    }
        }
    }
```

$$\frac{F(x,y)}{2^n} = \sqrt{\frac{\sum_{j=1}^n \frac{\prod_{i=1}^{10} x_{i,j}}{\prod_{i=1}^{15} frac}}{\sqrt{\sqrt{\int\int_a^b \frac{f(x)}{g(x)} dx}}}}$$

And even more is here:

```
    {
        F(x,y)
    \over
        2^n
    }
=
    \sqrt{
        {
```

```
                    \sum_{j=1}^n {
                                \prod_{i=1}^{10} x_{i,j}
                             \over
                                \prod_{i=1}^{15} {
                                                x_{i,j}
                                             \over
                                                y_{i,j}
                                             }
                             }
               \over
                    \sqrt{\sqrt{\int\!\!\int_a^b {
                                                f(x)
                                             \over
                                                g(x)
                                             }
                             dx}}
               }
        }
```

This leads to the following final output.

$$\frac{F(x,y)}{2^n} = \sqrt{\frac{\sum_{j=1}^n \frac{\prod_{i=1}^{10} x_{i,j}}{\prod_{i=1}^{15} \frac{x_{i,j}}{y_{i,j}}}}{\sqrt{\sqrt{\int\!\!\int_a^b \frac{f(x)}{g(x)} dx}}}}$$

14.9 Display Math Mode

This Section deals with some specific questions in relation to the display math mode. 14.10, p. 220, deals with multiline displays, which is also an issue specific to display math mode.

14.9.1 Vertical Glue and Penalty Parameters, Associated with Display Math Mode

There are certain penalties and glue parameters associated with display math mode. First we discuss the penalties:

1. `\predisplaypenalty` is the penalty associated with a (page) break before the display. It is set to 10000 in the plain format, because a formula on top of a page, not preceded by text, is undesirable.
2. The `\postdisplaypenalty` is the penalty inserted after a display formula. Its default is 0.

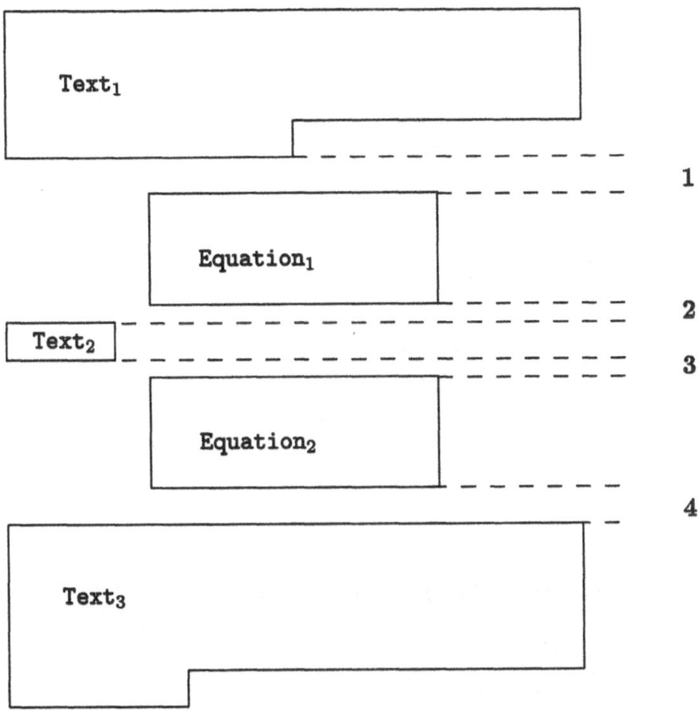

Figure 14.4. Illustration of **\abovedisplayskip**, **\abovedisplayshortskip**, **\belowdisplayskip** and **\belowdisplayshortskip**

Next are the following glue parameters (see Figure 14.4 on this page for a graphical illustration).

1. **\abovedisplayskip** and **\abovedisplayshortskip** are glues inserted before a math formula in display mode. The first glue is inserted, if the line preceding the formula extends into the left edge of the formula (case 1 in Fig. 14.4 on this page). Otherwise the other glue is inserted (case 3 of the same figure).

   The defaults in the plain format are **12pt plus 3pt minus 9pt** and **0pt plus 3pt** respectively.
2. Depending on which of the preceding glues was inserted, the formula is followed by **\belowdisplayskip** or **\belowdisplayshortskip** respectively. The defaults in the plain format are **12pt plus 3pt minus 9pt** and **7pt plus 3pt minus 4pt** respectively.

It is important to define those glues with stretchability and shrinkability. Besides **\parskip** the glue around equations is TEX's primary source to adjust the page length.

14.9.2 Equation Numbers, \eqno, \leqno

The way to enter equation numbers that appear on the *right hand margin* is as follows (notice the use of **\eqno**):

$$** ⟨formula⟩ **\eqno** ⟨formula⟩ **$$

Use **\leqno** to get numbers at the left hand margin. Equation numbers are set in text style. Here is a simple example:

```
$$
    X = \sum_{i=1}^{n} x_{i} \eqno (3)
$$
```

This input generated the following output:

$$X = \sum_{i=1}^{n} x_i \qquad (3)$$

On the other hand the following input

```
$$
    Y = \sum_{i=1}^{n} y_{i} \leqno (4)
$$
```

generates the following output:

$$(4) \qquad Y = \sum_{i=1}^{n} y_i$$

14.9.3 The Horizontal Positioning of Displayed Equations

By default displayed equations are centered between the left and the right margin of the text. Math mode does not take **\leftskip** and **\rightskip** into account. These glue values are only used by the line breaking algorithm.

Sometimes you want to have equations left flush, or indented from the left by a certain amount rather than having them right justified. Here is how this can be done.

First define macro **\LeftDisplay** (see the later following example for how this macro is applied):

$\mathcal{P}'$ • **leftdm.tip** •

```
\def\LeftDisplay #1$${%
```

\leftline prints its material left flush.

```
    \leftline{%
```

Not quite to the left: leave 20 pt distance from the left margin.

```
        \hskip 20pt
```

Start inline math mode, typeset the equation in display math style and end inline math mode.

```
            $
                \displaystyle {#1}
            $%
```

End `\leftline`. This macro will be invoked inside display math mode which must be ended.

```
            }%
        $$
}
```

• End of `leftdm.tip` •

Here is an example application of the preceding macro.

• `ex-leftdm.tip` •

```
\InputD{leftdm.tip}                     % 14.9.3, p. 218.
```

Make macro `\LeftDisplay` be invoked automatically at the beginning of every display math equation by assigning it to `\everydisplay`. When `\LeftDisplay` is invoked, `#1` of this macro becomes the complete math formula. Note that because the closing `$$` is used as delimiter of this macro's parameter (and therefore was thrown out by TeX when the parameter text is matched with the actual argument), the macro itself needs to reinsert `$$` back into the text to close off math mode.

```
\everydisplay = {\LeftDisplay}
$$
    X-Y = 8*12
$$
```

• End of `ex-leftdm.tip` •

The output generated by the preceding code reads as follows (the instruction to print the preceding example was enclosed in curly braces, because it is inappropriate to continue left flush displayed equations for the rest of this series).

$X - Y = 8 * 12$

14.9.4 More Display Mode Related Parameters, `\predisplaysize`, `\displaywidth` and `\displaysize`

There are three parameters which TeX uses for the typesetting of displayed equations.

1. `\predisplaysize`. Set to the effective width of the text line preceding the display.

2. **\displaywidth**. Set to the line width (that is **\hsize**). This value is not affected by the current value of **\leftskip** or **\rightskip**.
3. **\displayindent**. Set to the shift amount for line number **\prevgraf** + 2. Usually this value is zero, unless **\parshape** or hanging indentation is in effect. This value is not affected by the current value of **\leftskip** or **\rightskip**.

We glanced over many details in this chapter. For instance, the question arises why is there a parameter **\displaywidth**, if obviously **\hsize** could be used. See the TEXbook for details. Also note, that while these parameters are set at the beginning of a displayed equation, their values are not used until the end of the displayed equation is reached. In between the values can be changed by the user.

14.10 Multiline Displays

Next we discuss multiline displayed equation where more than one line belongs to the same mathematical equation.

14.10.1 Multiline Displays Without Equation Numbers

There are cases where a display extends over consecutive lines and one desires to align all equal signs. The **\eqalign** macro (which refers to **\halign**) provides this function. It is applied as follows:

```
\eqalign{  ⟨left-hand side_1⟩  &  ⟨right-hand side_1⟩\cr
           ⟨left-hand side_2⟩  &  ⟨right-hand side_2⟩\cr
           ⋮
           ⟨left-hand side_n⟩  &  ⟨right-hand side_n⟩\cr
}
```

We show an example here: the input

```
$$
    \eqalign{X_1, \ldots, X_n                          & = 0\cr
             Y_1, \ldots, Y_n, Y_{n+1}, \ldots Y_{2n} & = -1.\cr
            }
$$
```

generates:

$$\begin{aligned} X_1,\ldots,X_n &= 0\\ Y_1,\ldots,Y_n,Y_{n+1},\ldots Y_{2n} &= -1. \end{aligned}$$

You can use two `\eqaligns` in the same display—the two expressions will appear with their middle lines on the same height.

As with tables, you can insert vertical material using `\noalign` after any `\cr`. If we modify the previous example

```
$$
    \eqalign{X_1, \ldots, X_n                        & = 0\cr
                                                     \noalign{\vskip 5pt}
        Y_1, \ldots, Y_n, Y_{n+1}, \ldots Y_{2n} & = -1.\cr
    }
$$
```

then it generates the following output:

$$\eqalign{X_1, \ldots, X_n & = 0\cr \noalign{\vskip 5pt} Y_1, \ldots, Y_n, Y_{n+1}, \ldots Y_{2n} & = -1.\cr}$$

14.10.2 Multiline Displays With Equation Numbers

Next we discuss with multi-line displays with equation numbers. For that purpose the `\eqalignno` macro is defined. An optional equation number can be inserted before any `\cr` by writing "`&`⟨number⟩ `\cr`" instead. Thus there is a third column for equation numbers reserved.

The macro `\leqalignno` works the same way with the equation number placed on the left side of an equation.

14.10.3 Matrices

The printing of matrices is also based on TEX's `\halign`. For instance, macro `\matrix`, to be discussed shortly, uses `\halign` to do the real work.

Matrix elements are usually *centered.* If you want to justify them differently, use glue of type `\hfill`; see 38.3.14, p. IV-226, for an explanation.

14.10.3.1 Printing Matrices, `\matrix`

In order to type a matrix in TEX you need three things: the *left delimiter*, the *matrix elements*, and the *right delimiter*. Here is an example. The `\matrix` macro does *not* provide the parenthesis around the matrix elements. Those were generated by `\left(` and `\right)`. The following formula

$$A = \left(\matrix{x-\lambda & 1 & 0 \cr 0 & x-\lambda & 1 \cr 0 & 0 & x-\lambda \cr} \right)$$

has been typeset by typing

```
$$
A = \left( \matrix {x-\lambda&        1&            0\cr
                    0&                x-\lambda&    1\cr
                    0&                0&            x-\lambda\cr}
      \right)
$$
```

14.10.3.2 Matrices with Parentheses, `\pmatrix`

Because in general matrices are delimited by parentheses, there is a special macro `\pmatrix` which inserts the parentheses automatically. The preceding example therefore can be simplified to (generating the same output):

```
$$
    A = \pmatrix {x-\lambda&        1&            0\cr
                  0&                x-\lambda&    1\cr
                  0&                0&            x-\lambda\cr}
$$
```

The preceding source code generated the following output:

$$A = \begin{pmatrix} x-\lambda & 1 & 0 \\ 0 & x-\lambda & 1 \\ 0 & 0 & x-\lambda \end{pmatrix}$$

14.10.3.3 `\bordermatrix`

Sometimes there is a matrix with labels on top of its columns and left of its rows. The macro `\bordermatrix` is provided for this case. Its format is the same as the format of `\pmatrix`, but the parentheses are positioned differently: the opening parentheses is placed between the first and the second column and opening and closing parenthesis extend upwards only to the second row. Here is an example:

```
$$
Q = \bordermatrix {*&             1&            2\cr
                   A&             x-\lambda&    1\cr
                   B&             0&            x-\lambda\cr}
$$
```

$$Q = \bordermatrix{ * & 1 & 2 \cr A & x-\lambda & 1 \cr B & 0 & x-\lambda \cr}$$

14.10.3.4 More on Matrices

To insert dots into a matrix the instructions `\vdots` for *vertical* dots and `\ddots` for *diagonal* dots are provided. In addition, `\ldots` can be used for horizontal dots.

14.11 Token Parameters \everymath and \everydisplay

Along the lines of `\everyhbox` and `everyvbox` and other `\every...` token parameters, there are two token parameters which are automatically invoked by TeX:

- `\everymath` is evaluated at the beginning of every inline math equation.
- `\everydisplay` is evaluated at the beginning of every math display equation.

Similar to the discussion of 6.13, p. I-206, I decided to use these registers to count the number of math equations in this series (see 14.9.3, p. 218, for an application of `\everydisplay`). Here is the source code of a macro which does that.

$\mathcal{P}$ • `everyequ.tip` •

```
\newcount\EveryMathCount
\everymath = {\global\advance\EveryMathCount by 1 }
\newcount\EveryDisplayCount
\everydisplay = {\global\advance\EveryDisplayCount by 1 }
```

The following macro must be invoked at the end of every part of a document. This macro writes the number of math mode and display math equations to the log file.

```
\def\EndEveryMathDeal{%
    \wlog{}%
    \wlog{MATHCOUNT \the\EveryMathCount}%
    \wlog{DISPLAYCOUNT \the\EveryDisplayCount}%
    \wlog{}%
}
```

• End of `everyequ.tip` •

I then wrote a small command procedure which would collect these counter values from the log files of the various parts of this series and would add up all these values. The result of all this is the following: when processing this series 10175 inline and 457 displayed equations are generated.

Note that the preceding numbers are somewhat misleading: the macro `\angt` (`\angt{xx}` prints ⟨xx⟩), for instance, enters math mode twice to print $<$ and $>$. This, of course, is not really what math mode is all about. As far as displayed equations are concerned, the most frequent occurrence of display math mode in this series is to center entities such as paragraphs and tables (enclosed in vboxes). Those applications also don't really represent display math mode.

For an additional application of **\everydisplay** see 21.9.13, p. III-199.

14.12 Single and Double Math Shift Characters

If you tried to use display math mode in an **\hbox** as, for instance, in

```
    \hbox{$$\alpha$$}
```

you would probably be surprized about the errors generated by TEX. Let's try it. Here is the TEX source file used:

• ex-alpha.tip •

```
\nonstopmode
\hbox{$$\alpha$$}
\bye
```

• End of ex-alpha.tip •

The generated log file reads as follows:

• ex-alpha.log •

```
This is TeX, C Version 3.14 (...)
**&/usr/local/tex/lib/fmt/plain ex-alpha.tip
(ex-alpha.tip
! Missing $ inserted.
<inserted text>
$
<to be read again>
\alpha
l.2 \hbox{$$\alpha
$$}
I've inserted a begin-math/end-math symbol since I think
you left one out. Proceed, with fingers crossed.

! Extra }, or forgotten $.
<recently read> }

l.2 \hbox{$$\alpha$$}

I've deleted a group-closing symbol because it seems to be
spurious, as in `$x}$'. But perhaps the } is legitimate and
you forgot something else, as in `\hbox{$x}'. In such cases
the way to recover is to insert both the forgotten and the
deleted material, e.g., by typing `I$}'.

! Missing $ inserted.
<inserted text>
```

```
$
<to be read again>
\par
\bye ->\par
\vfill \supereject \end
l.3 \bye

I've inserted a begin-math/end-math symbol since I think
you left one out. Proceed, with fingers crossed.

! Missing } inserted.
<inserted text>
}
<to be read again>
\vfill
\bye ->\par \vfill
\supereject \end
l.3 \bye

I've inserted something that you may have forgotten.
(See the <inserted text> above.)
With luck, this will get me unwedged. But if you
really didn't forget anything, try typing `2' now; then
my insertion and my current dilemma will both disappear.

[1] )
Output written on ex-alpha.dvi (1 page, 248 bytes).
```

Here is what happened:

1. TEX regards display math mode inside an hbox as illegal. Therefore the first two `$$` will *not* be interpreted as "start display math mode," but as two separate math shift characters. The first one begins an inline math formula, and the second one ends it.
2. The `\alpha` is read in while TEX is in restricted horizontal mode and *not* in any math mode. TEX now generates an error and enters inline math mode.
3. The `$` after `\alpha` causes TEX to end this inline math mode.
4. The last `$` causes TEX to start inline math mode again.
5. When TEX subsequently sees the closing curly brace, TEX generates yet another error, and forces horizontal mode, before it terminates the equation.

It's an art to generate that many errors with that few tokens, isn't it?

14.13 Summary

In this chapter we learned the following:

- TeX builds a math list when it converts a math formula into boxes. This math list among other things contains atoms, style change instructions, generalized fractions and so forth. A math list is comparable to a syntax tree.
- TeX has alltogether eight styles. These styles are assigned after a math list was built, mainly for the purpose of determining what sizes to use to print various symbols.
- A math code consists of three parts: a class number (which classifies the symbol), a family number (which is responsible for selecting the proper font), and a character code (which defines the character code to be used within the specified family).
- A math code can be assigned to a character or control sequence using `\mathchardef` or `\mathcode`. Any math symbol can be printed using `\mathchar`.
- A math family consists of three fonts. Those fonts are identical (for all practical purposes) and differ only in size. Three families (family indices 0–2) are fixed. They must be assigned the math italics fonts, a math symbol font and a math extension font respectively.
- The plain format additionally defines five more families such as family 4 for roman text or symbols within math equations. The user can assign his or her own families using `\newfam`.
- There are two strategies for typesetting large equations. One is to define a sequence of macros, where each macro reflects a certain subformula based either on previously defined macros or on TeX's instructions for typesetting formulas. The other strategy is a top-down design whereby a formula is filled in a little at a time, in a top-down fashion, using a step-by-step refinement strategy.
- TeX inserts certain vertical glues before and after displayed equations. It also inserts implicit vertical glues before and after displayed equations.
- Multiline displays can be generated with `\eqalign`.
- Matrices can be generated very easily in TeX using `\matrix`, `\pmatrix` and `\bordermatrix`.
- The two token registers `\everymath` and `\everydisplay` are evaluated at the beginning of every inline or displayed equation.

15
Fonts in TEX

This chapter together with the next chapter discusses fonts in TEX. TEX comes with a standard set of fonts, called the Computer Modern fonts. Although our discussion will center around those fonts for the most part, the explanations of this and the following chapter should also allow you to add other fonts to TEX.

15.1 Magnification

As the word "magnification" suggests, it is possible to magnify a document. There are two different magnifications that should be discussed.

1. *Global magnification.* The global magnification is applied to the document as a whole. It applies to all pages. It cannot be changed in the middle of a document. The global magnification in TEX is like a photocopy machine which can reduce or enlarge a document.
2. *Font magnification.* Fonts can be magnified individually. For instance, you can load the Computer Modern Roman 10 pt font at 12 points and it will appear as a 12 point font (details follow shortly).

Effectively, looking at the final output, both magnifications multiply with each other: a specific character's size, as it is finally printed, is determined by the product of the font magnification of the font of that character, and the global magnification.

15.1.1 Specifying Magnifications in TEX

A magnification in TEX, in general, (that is a global as well as a font magnification), is specified as an *integer* value which is computed by multiplying the desired magnification by 1000 and rounding to the nearest integer. A magnification factor of 1000 (= 1.0 * 1000) therefore specifies a "neutral magnification." To double the dimensions in a document, you would have to use a magnification factor of 2000 (= 2.0 * 1000) (and you would have to have access to a printer that can print a document that large).

In this series I sometimes am not quite consistent in my terminology. I may talk about a magnification factor of 2.0 in which case I assume that you would really use the number 2000 when you actually programmed TEX. Or I might actually talk about magnification 2000 and by that mean a magnification by a factor of 2.0. There should be no problem in keeping things separate.

There are good reasons (to be explained later) to limit oneself to a standard set of magnifications. The list of those magnifications can be found in Table 16.1, p. 290. Besides "no magnification" (`\magstep 0`), there are six additional magnifications (`\magstephalf` and `\magstep 1...5`) available, which increase a document's dimensions and text by up to 150%.

15.1.2 Setting the Global Magnification, \mag

The global magnification is determined by assigning the appropriate integer value to `\mag`, a counter parameter. To double the size of a document one would have to issue the instruction `\mag = 2000`. The default, of course, is `\mag = 1000`.

As just mentioned, a magnification factor of 2000 would be a bad choice, because this value is not one from the standard set of magnifications. Therefore the usual procedure is to use the above-mentioned table and to look for the closest standard magnification, `\magstep 4` in that case.

Note that writing `\magstep 4` alone (or any of the other standard magnifications) does *not* change the magnification, because `\magstep 4` simply expands to 2073, that is writing `\magstep 4` simply prints 2073. All `\magstep` control sequences are simply abbreviations for certain numbers (see column 3 in Table 16.1, p. 290). You actually need to write `\mag = \magstep 4`, that is assign the magnification value to `\mag`.

Setting the global magnification by assigning a value to `\mag` should be done as early as possible in a document as one of the very first instructions in the document. Actually it can be done anywhere as long as two conditions are met:

- The *first* page was not yet written to the `dvi` file.
- No `true` dimension was used (see 15.1.3 on the next page).

The effect of setting the global magnification of a document to a value which is different from 1 is similar to using a photocopy machine with an enlargement or reduction option: the document's layout is *not* affected by the global magnification, the line breaks and page breaks stay the same regardless of the chosen magnification. The document is simply printed at a different size (and if you print an enlarged document your printer and paper had better be large enough).

15.1.3 Unmagnified or `true` Dimensions

The global magnification of a document also magnifies every dimension in this document. For instance, if one specifies some vertical space, then this vertical space is also magnified.

Sometimes though one needs to specify a distance in which TEX is *not* supposed to apply the magnification, regardless of the global magnification in effect. For instance, one might want to glue a diagram into a document, but this diagram will not be magnified and therefore the space reserved for it should always be the same, regardless of the global magnification.

One thing one can do is account for the global magnification and to *divide* that dimension by the magnification factor. Then, when the document is magnified, the dimensions come out as desired.

This is cumbersome and there is an easier approach. TEX has, so-called, *true dimensions* where TEX automatically compensates for the global magnification by simply adding the word `true` to the dimension unit chosen. For instance, writing `\hskip 2.0true in` will cause TEX to skip horizontally 2.0 in regardless of the current global magnification. The `true` specification works with any of TEX's dimension units.

15.1.4 `\magnification`

Let me now show another form of changing the magnification of a document. In this case, the document is printed at a larger size due to the magnification, but the original page sizes are preserved. Thus the default values for the horizontal size (`\hsize`) and vertical size (`\vsize`) stay in effect.

One way to achieve this effect is to systematically replace all fonts by bigger fonts. Note that because `\hsize` and `\vsize` remain the same the layout of the document changes completely. For instance, if the document is enlarged, it will usually become longer (more pages).

However there is an easier way to do all of this. First change the magnification, then set the values of `\hsize` and `\vsize` to their defaults using "true dimensions:"

```
\mag = \magstep 1
\hsize = 6.5 true in
\vsize = 8.9 true in
```

TEX provides the macro `\magnification` to execute the three TEX lines above. For instance, the above example could have been abbreviated to `\magnification = \magstep 1` (note: the desired magnification is *not* hardwired into the macro but must be provided in the way just outlined).

The `\magnification` macro is defined as follows (it uses `\afterassignment` discussed in 23.1, p. III-235:

```
\catcode`@ = 11
```

```
\def\magnification{%
    \afterassignment\m@g\count@
}

\def\m@g{%
    \mag = \count@
    \hsize = 6.5 true in
    \vsize = 8.9 true in
    \dimen\footins = 8 true in
}
\catcode`@ = 12
```

The next Section deals with the loading of fonts, where you will also find a discussion of how a font is loaded when magnified (so far only global magnification was dealt with).

15.2 The Basics of the Handling of Fonts by TEX

TEX retrieves all the necessary font information from `tfm` files. "Loading a font" means TEX is reading in a `tfm` file.

Note that initially TEX does *not* have any font information loaded (in particular, `initex` and `virtex` have no font information loaded; see 17.1.1, p. 313). Normally a format loads a certain set of standard fonts ("standard" as defined by the particular format) and the user may or may not load some additional fonts on his or her own. The font loading process is the same in both cases however.

15.2.1 `\nullfont`

If no font has been loaded, then the null font (one can invoke it by the primitive `\nullfont`) is in effect. This font has no characters, so of course you usually won't use this instruction.

15.2.2 Character Codes

Characters are administered internally in the TEX program as numbers. These numbers are based on the ASCII code and are referred to as *character codes.* TEX uses an 8 bit ASCII code and allows for the direct access of 256 characters with character codes 0...255 (this is only since the arrival of TEX 3.0; before that the maximum number of directly accessible character codes was 128 (7-bit ASCII)).

Character codes up to 255 are legal in TEX, but for codes in the range of 128...255, the `\char` instruction must be used to access those characters. To find out more about character codes see 16.6.1, p. 296.

15.2.3 Characters Are Horizontal Boxes in TEX

For TEX, each character is simply a horizontal box (or hbox for short). Fig. 6.1, p. I-162, shows a "TEX box." A box in TEX is an entity which has a certain *height*, *depth* and *width*. A box also has a *reference point* and a *baseline*. The three dimensions of a character box naturally depend on the dimensions of a character and the font; a character box is about the size (a little larger frequently) of the "minimum bounding box of the character." Note that TEX has no access to information about the *shape* of a character, and TEX does not even need this information. When TEX typesets text, all it needs to do is *reserve the proper amount of space* for each character, and for that purpose, the dimensions of character boxes are totally sufficient.

If a word like "Sample" is printed by TEX, then the six character boxes of this word will be lined up the same way TEX lines up hboxes: the baselines of all boxes are on the same height and the boxes are adjacent to each other. Look at the following diagram, where the word "Sample" is printed in three different ways:

1. First the word "Sample" is printed simply as it would appear in some text.
2. Then the word "Sample" is printed where, in addition to each character, the character's box is also printed.
3. Finally the word "Sample" is "printed" where only character boxes are printed; this output comes closest to TEX's "mental picture" of characters.

Here is the complete source code of this example, followed by the output. First we define a macro `\hboxE` which

$\mathcal{P}'$ • `charbo.tip` •

```
\InputD{box-mac.tip}                        % 9.3.14, p. I-343.
```

Define a macro that draws an empty hbox of the size determined by character #1.

```
\def\hboxE #1{%
     {%
```

Get the character into box register 0.

```
          \setbox0 = \hbox{#1}%
```

Load an empty box into box register 1 and then transfer all dimensions from box register 0 (that is the dimensions of this character) to box register 1.

```
          \setbox1 = \hbox{}%
          \wd1 = \wd0
          \ht1 = \ht0
```

```
        \dp1 = \dp0
```

Print an empty box with the "right" dimensions.

```
        \HboxR{\box 1}%
    }%
}
```

• End of charbo.tip •

• sample-cb.tip •

```
\InputD{charbo.tip}                    % 15.2.3, p. 231.
```

Now print the word three times, in three different ways.

```
$$
\vbox{%
    \Huge
    \hbox{Sample}%
    \hbox{\HboxR{S}\HboxR{a}\HboxR{m}\HboxR{p}\HboxR{l}\HboxR{e}}%
    \hbox{\hboxE{S}\hboxE{a}\hboxE{m}\hboxE{p}\hboxE{l}\hboxE{e}}%
}
$$
```

• End of sample-cb.tip •

The output generated by the preceding source code reads as follows:

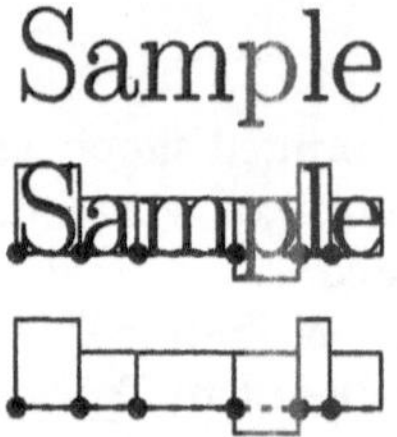

15.2.4 Determining the Sizes of Characters in a Font, \ReportCharSize

The next macro, when given the character code of a character, prints the size of the character into the log file. This is done by storing the character in a box register and then writing the dimensions of this box register. If you have read the preceding macro definition, then the macro definition of this Subsection should be straightforward to understand.

The macro `\ReportCharSize` has two parameters:

- `#1`. The font to which the character belongs.
- `#2`. The character code (*not* the character) of the character of which the dimensions should be written to the log file.

$\mathcal{P}'$ • `fo-char.tip` •

```
\def\ReportCharSize #1#2{%
```

Start a group, save character code in counter register 0. Load box register 0 with the character (switch to the proper font first).

```
    {%
        \count0 = #2\relax
        \setbox 0 = \hbox{#1\char\count0}%
```

Now use `\wlog` to write the requested information to the log file.

```
        \wlog{\string\ReportCharSize: Font \string#1,
            character code \the\count0}%
        \wlog{ht / dp / wd: \the\ht0 \space / \the\dp0
            \space / \the\wd0}%
    }%
}
```

• End of `fo-char.tip` •

Here is a sample application of this macro that writes the sizes of characters "A" and "e" (typewriter-like font `\tt`) to the log file. The reprint of the log file shows that the two characters have the same width as we expect from a fixed spaced font.

Here is the source file used for this example:

• `ex-fo-char.tip` •

```
\input inputd.tip
\InputD{fo-char.tip}                          % 15.2.4, p. 233.
\ReportCharSize{\tt}{`\A}
\ReportCharSize{\tt}{`\e}
\ReportCharSize{\rm}{`\A}
\ReportCharSize{\rm}{`\e}
\bye
```

• End of `ex-fo-char.tip` •

And here is the resulting log file generated by the above source code.

• `ex-fo-char.log` •

```
This is TeX, C Version 3.14 (...)
**&/usr/local/tex/lib/fmt/plain ex-fo-char.tip
(ex-fo-char.tip (inputd.tip
(namedef.tip
) (inputdl.tip
\@InputDStream=\write0
))
(fo-char.tip
)
\ReportCharSize: Font \tt, character code 65
ht / dp / wd: 6.11111pt / 0.0pt / 5.24995pt
\ReportCharSize: Font \tt, character code 101
ht / dp / wd: 4.30554pt / 0.0pt / 5.24995pt
\ReportCharSize: Font \rm, character code 65
```

```
ht / dp / wd: 6.83331pt / 0.0pt / 7.50002pt
\ReportCharSize: Font \rm, character code 101
ht / dp / wd: 4.30554pt / 0.0pt / 4.44444pt
)
No pages of output.
```

15.2.5 A Macro to Print the Character Sizes of All Characters of a Particular Font, `\TfmSizeTable`

The macro `\TfmSizeTable` generates a table with the dimensions of selected or all characters of a particular font. Here is the related source code:

• tfmsizes.tip •

```
\InputD{oct.tip}                          % 27.1.4.3, p. III-406.
\InputD{endrec.tip}                       % 27.1.2, p. III-399.
\catcode`\@ = 11
```

Allocate a counter register for the "current" character code of the following macro.

```
\newcount\@TfmSizesCharacterCode
```

Allocate counter register which specifies the character code up to which the table is printed.

```
\newcount\@TfmSizesLimit
```

A private box register is needed too.

```
\newbox\@TfmSizesBox
```

The definition of macro `\TfmSizeTable` begins here. This macro has the following three parameters:

- `#1`. A complete "font loading specification" which will be used with `\font` to load a particular font, as, for instance, "`cmr10 scaled \magstep 2`."
- `#2`. The minimum character code at which point the font size table will start. Typically 0.
- `#3`. The maximum character code up to which the table will go. Typically 127 (255 for fonts with 256 characters).

When the above macro is called, it generates a table which contains the size of every character in a particular font. Each character's code is printed as octal number (first column of the generated table), each character is actually printed (second column) and then the height, depth and width (third, fourth and fifth column of the generated table) of each character are listed.

```
\def\TfmSizeTable #1#2#3{%
```

Save all parameters.

```
        \font\@TfmSizesFont = #1
        \@TfmSizesCharacterCode = #2
        \@TfmSizesLimit = #3
        \begingroup
```

Start the table here. Columns are as described above.

```
                                    \tabskip = 20pt
        \halign\bgroup
            \tt'\Oct{##}:\hfil&
            \hfil##\hfil&
            \hfil##&
            \hfil##&
            \hfil##%
        \cr
```

Now start the loop which actually generates the table.

```
        \@TfmSizesLoop
    }
```

The following instructions are executed inside a **\noalign**. This prevents TEX from starting the first column and expanding the template of the first column.

The following macro uses the standard method of recursion, where either of two macros is executed resulting in a continuation or a termination of the recursion; see 27.1, p. III-397, for details on recursion.

```
    \def\@TfmSizesLoop{%
        \noalign\bgroup
```

The recursion goes one character beyond the highest character code for which we need to print a row in the table. The table is terminated once this has happened.

```
        \RecursionMacroEnd{%
            \ifnum\@TfmSizesCharacterCode > \@TfmSizesLimit}%
            {\@TfmSizesDone}{\@TfmSizesOneLine}%
    }
```

The following macro prints one line in the table which contains the character code, the character itself, and its dimensions.

```
    \def\@TfmSizesOneLine{%
        \global\setbox\@TfmSizesBox =
            \hbox{\@TfmSizesFont\char\@TfmSizesCharacterCode}%
```

Next terminate the **\noalign\bgroup**.

```
        \egroup
        \the\@TfmSizesCharacterCode&
        \copy\@TfmSizesBox&
        \the\ht\@TfmSizesBox&
        \the\dp\@TfmSizesBox&
        \the\wd\@TfmSizesBox
        \global\advance\@TfmSizesCharacterCode by 1
```

The end of the row is reached.

```
        \cr
```

Table 15.1. Character sizes of characters a–k of the Computer Modern roman 10 pt font.

'141:	a	4.30554pt	0.0pt	5.00002pt
'142:	b	6.94444pt	0.0pt	5.55557pt
'143:	c	4.30554pt	0.0pt	4.44444pt
'144:	d	6.94444pt	0.0pt	5.55557pt
'145:	e	4.30554pt	0.0pt	4.44444pt
'146:	f	6.94444pt	0.0pt	3.05557pt
'147:	g	4.30554pt	1.94444pt	5.00002pt
'150:	h	6.94444pt	0.0pt	5.55557pt
'151:	i	6.67859pt	0.0pt	2.77779pt
'152:	j	6.67859pt	1.94444pt	3.05557pt
'153:	k	6.94444pt	0.0pt	5.2778pt

Restart the recursion.

```
        \@TfmSizesLoop
}
```

The following macro finishes off everything (it is called after the last table row was printed). First, `\noalign` is terminated (which really did not do anything), then the `\halign`, and finally the group (started by `\begingroup`) is ended.

```
\def\@TfmSizesDone{%
        \egroup
        \egroup
        \endgroup
}
\catcode`\@ = 12
```

• End of `tfmsizes.tip` •

An example application of the preceding macro can be found in Table 15.1 on this page where the sizes of characters a–k of the Computer Modern Roman 10 pt font are displayed. The `\TfmSizeTable` macro to print this table was invoked as follows:

```
$$
        \vbox{
            \TfmSizeTable{cmr10}{`\a}{`\k}
        }
$$
```

15.2.6 Determining the Length of a String, `\StringLength`

To compute the length of a string, the macro `\StringLength` will make use of the fact that *all* characters in a fixed-width font have the same width. The macro `\StringLength` has one parameter, `#1`, a string (or a macro expanding to a string). After the macro's execution the counter register, `\StringLengthResult`, contains the number of characters stored in the string. The string itself is not printed or changed in any way.

Here is the source code of this macro.

$\mathcal{P}'$ • **strleng.tip** •

```
\newcount\StringLengthResult
\catcode`\@ = 11
```

Two "private" box registers are allocated.

```
\newbox\@StringLengthBoxA
\newbox\@StringLengthBoxB
```

The definition of the macro `\StringLength` starts here.

```
\def\StringLength #1{%
```

Compute the width of the whole string and the width of one character.

```
     \setbox\@StringLengthBoxA = \hbox{\tt #1}%
     \setbox\@StringLengthBoxB = \hbox{\tt A}%
```

Now divide the width of the whole string by the width of one character, which is the length of the string.

```
     \StringLengthResult = \wd\@StringLengthBoxA
     \divide\StringLengthResult by \wd\@StringLengthBoxB
}
\catcode`\@ = 12
```

• End of `strleng.tip` •

Here is an example application of this macro. The following source code is used:

• **ex-strleng.tip** •

```
\input inputd.tip
\InputD{strleng.tip}                        % 15.2.6, p. 237.
```

Save a string in a macro and compute its width.

```
\def\MyString{ABCDEF-890}
\StringLength{\MyString}
```

Now print some text including the width of the string.

```
\centerline{%
     The string ``{\tt\MyString}'' has
     \the\StringLengthResult\space characters.%
}
```

• End of `ex-strleng.tip` •

Executing this TEX source code prints the following text:

The string "**ABCDEF-890**" has 10 characters.

15.2.7 The Loading of Fonts by Loading TFM Files

A font in TEX is usually labeled in one of the following three ways.

1. Recall that the font has a "full name" such as "Computer Modern roman 10 pt" (the term "Computer Modern fonts" will be explained shortly).
2. From this font name a (much shorter) name such as `cmr10` (Computer Modern roman 10pt) is derived. This name is used for two purposes:
   (a) To name the `tfm` file of the font. For instance, the `tfm` file of our example font would be called `cmr10.tfm`. Obviously the file extension `tfm` stands for "TEX font metric."
   (b) To name the pixel files of this font such as `cmr10.300gf`, `cmr10.360gf` and so forth, all to be discussed later.

   Most of the `tfm` file names of the Computer Modern fonts are formed along the same lines as the name `cmr10`.
3. In TEX itself, to actually switch to a particular font, a *control sequence* is used. For instance, `\cmrten` may be defined in such a way that it causes TEX to switch to Computer Modern Roman 10 pt. Shortly it will be discussed how one can get TEX to associate a font with a control sequence such as `\cmrten`.

15.2.8 The Information Stored in TFM Files

A `tfm` filke stores a great deal of information. I will not list all the information contained in such files (see Knuth (1986b), section 539, for details):

1. The width, height and depth of each character. As pointed out before, TEX does not have any information about the shapes of the characters of a font.
2. Kerning and ligature information (see 5.4.8, p. I-140).
3. The default interword spacing and spacing after a punctuation symbol at the end of sentences (see 15.2.15, p. 243). This is the amount of horizontal space inserted between words if this particular font is active; see 16.2.4, p. 277, for complete details.
4. The design size of the font plus other font parameters.
5. A checksum to ensure consistency among the `tfm` files used by TEX and the pixel files of the driver used to print a document; see 17.8, p. 323, for details.

15.2.9 The Instruction \font Loads a Font

Assume you want to load the Computer Modern roman 10 pt font. The `tfm` file is called `cmr10.tfm`, and TEX's `\font` instruction would be used to load this font.

You have to give the font a "control sequence name" which you will use to switch to this font once it was loaded. For the example I chose `\cmrten`. Observe that `\cmr10` is *not* a legal control word because it contains numbers, and therefore cannot be used as control sequence name for this font.

The example font would be loaded the following way:

```
\font\cmrten = cmr10
```

Observe that the `\font` instruction only *loads* the specified font. It does *not* cause TEX to actually switch to this font.

Later we will also discuss loading *magnified* fonts. See 15.2.11 on this page for details. The question of how TEX goes about *locating* `tfm` files is discussed in 17.4, p. 318.

15.2.10 Changing Fonts Using Grouping

In general, 99% of a document is set in the document's "base font." In this series I use the Computer Modern roman 10 pt font, which is the most commonly used font with TEX. Any font which is different from the base font of a document is normally only used for a few words at a time. For instance, *italic* might be used to emphasize a word or two, or **boldface** may be used in a heading.

It is natural to use *grouping* for font changes as follows (grouping is explained in more detail in 19.4.2, p. III-100):

1. A group is started by an opening curly brace right before a font change. This group instructs TEX to remember the currently active font.
2. Next follows an instruction to change to the new font, like `\it` to print the following text in italic.
3. This font change instruction is followed by the text to be printed in the new font.
4. Finally a closing curly brace is entered, which tells TEX to terminate the previously started group, thereby reverting to the previous font.

Here is an example: to print a word in *italic* and subsequently one in **boldface**, the following text was entered:

```
... to print a word in {\it italic\/} and then
one in {\bf boldface}, the following text ...
```

The italic correction "\/" is discussed in 15.3.6, p. 256.

15.2.11 Loading Fonts Magnified, `\font \xx = cm... scaled`

It is *not* possible to magnify an already loaded font or to change the magnification of a font "on the fly," that is, while the font is being used. When a particular font is loaded at two different magnifications, then TEX looks at these fonts as two separate fonts.

To load a font magnified, simply append **scaled** followed by the desired magnification to the regular **\font** instruction. Here is an example where the Computer Modern roman 10 pt font magnified by 1.44 is loaded. This leads to approximately a 14 pt font (14.44 pt is the precise value) and hence the choice of the name **\fourteenrm**.

```
\font\fourteenrm = cmr10 scaled \magstep 2
```

The syntax of the **\font** command is such that you could use any number for a magnification factor; as discussed before one usually should restrict oneself to one of the standard **\magstep**... magnification factors.

The following examples give you a visual idea about the results of font magnification. The same font (Computer Modern roman 10 pt) is printed at six different magnifications. Here is the source code of an example:

• **ex-fomag.tip** •

```
\InputD{chboxd.tip}                          % 4.5.10, p. I-102.
```

First the necessary fonts are loaded.

```
\font\cmrmagzero =    cmr10 scaled \magstep 0
\font\cmrmaghalf =    cmr10 scaled \magstephalf
\font\cmrmagone =     cmr10 scaled \magstep 1
\font\cmrmagtwo =     cmr10 scaled \magstep 2
\font\cmrmagthree =   cmr10 scaled \magstep 3
\font\cmrmagfour =    cmr10 scaled \magstep 4
\font\cmrmagfive =    cmr10 scaled \magstep 5
```

Define the macro **\LB** which has one argument, **#1**, which is some text. This text is saved inside a box register, then this box register is printed with slightly increased dimensions.

```
\def\LB #1{%
    \setbox0 = \hbox{#1}%
    \AdvanceBoxDimension{\ht0}{4.0pt}%
    \AdvanceBoxDimension{\dp0}{4.0pt}%
    \box0
}
```

Start display math mode, start a vbox, start a table.

```
$$
    \vbox{
        \offinterlineskip
        \hrule
                                              \tabskip = 0pt
        \halign{
```

The first column contains a strut.

```
            \vrule height 10pt depth 5pt width 0pt#&            % 1
```

The second column, fourth columns, etc., contain vertical rules. The columns inbetween (there are four of those) contain the printing of the sample text, etc.

```
            \vrule#\relax                 \tabskip = 15pt&    % 2
            \hfil#\hfil&                                      % 3
            \vrule#&                                          % 4
            \hfil#\hfil&                                      % 5
            \vrule#&                                          % 6
            \hfil#\hfil&                                      % 7
            \vrule#&                                          % 8
            \hfil#\hfil&                                      % 9
            \vrule#\relax                 \tabskip = 0pt      % 10
```

End of the table preamble.

```
        \cr
```

First set of subheadings.

```
            &&{\tt\string\magstep\space 0}&&
            {\tt\string\magstephalf}&&
            {\tt\string\magstep\space 1}&&
            {\tt\string\magstep\space 2}&
        \cr
```

First set of samples.

```
            &&\cmrmagzero\LB{Age}&&
              \cmrmaghalf\LB{Age}&&
              \cmrmagone\LB{Age}&&
              \cmrmagtwo\LB{Age}&
```

End of the first set of samples. Draw a horizontal rule and have the second set.

```
        \cr\noalign{\hrule}
            &&{\tt\string\magstep\space 3}&&
              {\tt\string\magstep\space 4}&&
              {\tt\string\magstep\space 5}&&
            &
        \cr
            &&\cmrmagthree\LB{Age}&&
              \cmrmagfour\LB{Age}&&
              \cmrmagfive\LB{Age}&&
            &
        \cr
```

End of table, end of vbox, end of display mathmode.

```
        }
        \hrule
    }
$$
\vskip -12pt
```

• End of `ex-fomag.tip` •

The output generated by the preceding source code appears in Figure 15.1 on the next page.

From a font designer's point of view, font magnification is something *undesirable* to do, because fonts *do not* scale proportionally! A 20 pt font is *not*

\magstep 0	\magstephalf	\magstep 1	\magstep 2
Age	Age	Age	Age
\magstep 3	\magstep 4	\magstep 5	
Age	Age	Age	

Figure 15.1. Font magnification example.

twice the size of the 10 pt font (see page 16 of the TeXbook for an example). In other words, you should load fonts unmagnified, if you can. Therefore, instead of loading `cmr10 scaled \magstep 1`, load `cmr12`, if you can (otherwise run METAFONT so you can load `cmr12` afterwards). A magnified font is intended for the photographic reduction of a document.

15.2.12 Loading Fonts Magnified Using "`at xpt`"

There is another form of the `\font` instruction to load a magnified font, which is very similar to the `scaled` based approach discussed in the preceding Subsection. In this approach, you use the keyword `at` instead of `scaled`. The keyword `at` must be followed by the size of the font at which the font is to be used. Here is an example. The following instruction loads a 12 pt font.

```
\font\xx = cmbx10 at 12pt
```

Because the basefont is a 10 pt font, the following instruction loads the same font:

```
\font\xx = cmbx10 scaled \magstep 1
```

15.2.13 Looking at Global and Font Magnification Together

Remember as far as the final output is concerned, global and font magnification multiply with each other. Look at the following example:

```
\mag = \magstep 1
\font\cmrseventeen = cmr10 scaled \magstep 3
```

When `\cmrseventeen` is invoked, the final result will print as a 21 pt font approximately. The global and the font magnification multiplied with each other correspond to a magnification factor of 2.074 (`\magstep 4`).

The magnification factors form a geometric series, that is

`\magstep` $i *$ `\magstep` $j =$ `\magstep` $(i + j)$

Table 15.2. A font magnification table.

Base size	\magstep						
	0	0.5	1	2	3	4	5
5	5.00	5.47	6.00	7.20	8.64	10.37	12.44
6	6.00	6.57	7.20	8.64	10.37	12.44	14.93
7	7.00	7.67	8.40	10.08	12.10	14.52	17.42
8	8.00	8.76	9.60	11.52	13.82	16.59	19.91
9	9.00	9.86	10.80	12.96	15.55	18.66	22.39
10	10.00	10.95	12.00	14.40	17.28	20.74	24.88
12	12.00	13.14	14.40	17.28	20.74	24.88	29.86
17	17.00	18.61	20.40	24.48	29.38	35.25	42.30

This font table is nothing else than a multiplication table. For instance, a Computer Modern roman 8 pt font at magnification `\magstep2` results in a 11.52 pt font (row "8", column "2").

15.2.14 A Font Magnification Table

Table 15.2 on this page contains a font magnification table. It shows the resulting font size given a basic font size and the magnification applied to this font. This table comes in handy, if one needs to generate a specific font at a specific size but doesn't have the font available at the desired size and therefore must resort to magnification.

By the way, this table was generated with the help of a short C program which generates TEX instructions to typeset the table directly.

15.2.15 Font Dimension Parameters, `\fontdimen`

When TEX loads a `tfm` file, this `tfm` file contains *font dimension parameters* which specify certain dimensions within a font. These parameters are discussed now.

The instruction format of `\fontdimen` is `\fontdimen` ⟨parameter number⟩ ⟨font⟩, that is, for instance, `\the\fontdimen 2 \tenrm` prints the second font dimension parameter of Computer Modern roman 10 pt.

There are a maximum of 24 font parameters, of which parameters 8 to 24 are for math mode dimensions. The meaning of the first seven parameters is the following:

- `\fontdimen 1`. This parameter specifies the slant of the font per pt. This information, which says "how far characters are leaning over" (usually to the right), is necessary for the proper placements of accents.
- `\fontdimen 2`. This parameter specifies the default interword space (see 16.2, p. 275, for details).
- `\fontdimen 3`. This parameter specifies the stretchability of the default interword space (see 16.2, p. 275, for details).

- `\fontdimen 4`. This parameter specifies the shrinkability of the default interword space (see 16.2, p. 275, for details).
- `\fontdimen 5`. This parameter specifies the x-height of characters of that font (1 ex in this font is as much as this parameter specified).
- `\fontdimen 6`. This parameter specifies the quad width when this font is used (1 em in this font is as wide as this parameter specifies). This parameter, for regular fonts (that is not boldface, italic, slanted, or modified in any other way), is usually identical with the font size.
- `\fontdimen 7`. This parameter specifies the size of *extra space*, the space which can be additionally inserted after characters ending a sentence (such as a period); see 16.2, p. 275, for details.

• fodimen-tab.tip •

The following macro source code is loaded so dimensions below can be printed as "12.34 pt" instead of "12.34pt" (note the space between value and dimension unit).

```
\InputD{droppt.tip}                          % 18.1.11, p. III-12.
```

The following four font loading instructions are repeated here for the sake of clarity (they are already part of the plain format and so it is not really necessary to repeat these instructions).

```
\font\tenrm = cmr10
\font\tenit = cmti10
\font\tenbf = cmbx10
\font\tentt = cmtt10
\def\ttstring{\tt\string}
\def\FontDimenTwo{\expandafter\DropPoints\the\fontdimen2}
\def\FontDimenThree{\expandafter\DropPoints\the\fontdimen3}
\def\FontDimenFour{\expandafter\DropPoints\the\fontdimen4}
\def\FontDimenSeven{\expandafter\DropPoints\the\fontdimen7}
```

Here a table is built using `\halign`.

```
                                    \tabskip = 0pt
\halign{%
    \gdef\TableFont{#}%
    \expandafter\ttstring\TableFont \tabskip = 10pt&
                                          % 1. font used.
    #\expandafter\FontDimenTwo\TableFont
        \DropPointsResult\space pt\hfil&  % 2. \fontdimen 2
    #\expandafter\FontDimenThree\TableFont
        \DropPointsResult\space pt\hfil&  % 3. \fontdimen 3
    #\expandafter\FontDimenFour\TableFont
        \DropPointsResult\space pt\hfil&  % 4. \fontdimen 4
    #\expandafter\FontDimenSeven\TableFont
        \DropPointsResult\space pt\hfil   % 5. \fontdimen 7
                                    \tabskip = 0pt
\cr
```

Create the heading.

```
    \omit&
    \omit\hfil{\tt \string\fontdimen 2}\hfil&
    \omit\hfil{\tt \string\fontdimen 3}\hfil&
    \omit\hfil{\tt \string\fontdimen 4}\hfil&
    \omit\hfil{\tt \string\fontdimen 7}\hfil\cr
    \tenrm&&&&\cr
    \tenbf&&&&\cr
    \tenit&&&&\cr
    \tentt&&&&\cr
}
```

• End of fodimen-tab.tip •

The table generated by the above source code reads as follows:

	\fontdimen2	\fontdimen3	\fontdimen4	\fontdimen7
\tenrm	3.33333 pt	1.66666 pt	1.11111 pt	1.11111 pt
\tenbf	3.83331 pt	1.91666 pt	1.27777 pt	1.27777 pt
\tenit	3.57774 pt	1.53333 pt	1.0222 pt	1.0222 pt
\tentt	5.24995 pt	0.0 pt	0.0 pt	5.24995 pt

15.2.16 The \fontname Instruction

The **\fontname** instruction is, in a certain sense, the inverse instruction of **\font**. It must be followed by the control sequence which switches to a particular font. It generates the name of the **tfm** file for this font. For instance, after **\font\xx = cmr10**, **\fontname\xx** prints **cmr10**. If a font were loaded magnified (for instance, **\font\yy = cmr10 scaled \magstep 2**), **\fontname\yy** prints the font file name and size the font is used as, that is **cmr10 at 14.4pt** in the preceding example.

15.3 The Fonts of TEX

15.3.1 Some Basic Typesetting Terminology

Let me first introduce some typesetting terminology.

1. *Proportionally spaced fonts* have some characters with different widths As previously mentioned, the text you are currently reading is typeset in a *Computer Modern roman 10 point* font (**\rm**). This font is proportionally spaced. For example, the widths of a lowercase "i" and of an uppercase "W" of this font are clearly different. This does *not* mean that there could not be characters which have the same width. For instance, "U" and "V" in this font have the same widths.

2. *Mono-spaced fonts* have characters which all have the same width. For instance, the *Computer Modern Typewriter font* (`\tt`) is such a font. In this font therefore all characters have the same widths, and so do in particular "`i`" and "`W`."
3. *Left flush and right flush.* Text generated by a typesetting system is usually set left and right flush. This is also TEX's default. Observe that the necessary adjustments to the line lengths of lines to achieve such left and right flush margins is done by adjusting the *interword space*, the space between words, by small amounts. Naturally, the smaller these adjustments are, the better is the layout quality of some paragraphs (TEX's algorithm actually takes additional factors into account).
4. *Ragged right.* Ragged right text means that different lines have different length. The space between words is fixed (i.e., the same for each line) because there is no need to adjust the line length.
5. *Interword space.* The interword space is the space between the words of the text of a paragraph. Interword space is an important issue in this chapter as well as the preceding chapters on paragraphs.

   Two remarks about interword space:

   (a) When using proportionally spaced fonts in TEX, the interword space is normally allowed to stretch and shrink to allow for the adjustments of line lengths in a left flush and right flush type of page layout. The term "normally" here refers to using the standard fonts of TEX, the Computer Modern fonts.

   (b) In mono-spaced fonts the interword space is normally not adjusted (the definition of "normally" is the same as above). Therefore, text typeset in mono-spaced fonts normally uses a ragged right type of text layout.

   We have so far discussed the default setups: ragged right and mono-spaced fonts, left and right flush, and proportionally spaced fonts. This Chapter will explain how you can set up any font in any way you want (for instance, using a fixed spaced font with left and right-justified text).
6. Normally the space between a punctuation symbol ending a sentence and the first word of the following sentence is larger than the regular interword space; it is therefore called *extended.*

   If, on the other hand, *French spacing* is in effect, then such a space is *not* extended, but regular. Details will follow shortly.

15.3.2 METAFONT and the Computer Modern Fonts

METAFONT is the companion program of TEX. The basic idea of generating fonts with METAFONT is as follows: the shapes of characters are described in terms of pen movements in a coordinate system. From this information METAFONT generates the `tfm` and `gf` files of a font. TEX needs the `tfm` file for a font to have the measurements of this font, and the `gf` file of this font is used by the driver to print the text in this font (more details will follow shortly).

The *Computer Modern fonts* are a special set of fonts and were specifically designed for TEX. They are also used in this series. Other fonts can be used with TEX too as will be discussed shortly.

15.3.3 An Overview of Font Types Used With TEX

The following font classification can be used with TEX.

1. *Computer Modern text fonts.* The fonts within the Computer Modern fonts are used to print regular text. These fonts are the most commonly used among the Computer Modern fonts. Because these fonts belong to the Computer Modern fonts, they are METAFONT-based.
2. POSTSCRIPT *text fonts.* The fonts of the page description language POSTSCRIPT. See 16.10.1, p. 309, for the relationship of these fonts to METAFONT-based fonts.
3. *Third party text fonts.* Fonts provided by another vendor; see 16.10.2, p. 310, for a discussion of how third party fonts can be used in TEX.
4. *Computer Modern math fonts.* Fonts of the Computer Modern fonts needed for the typesetting of mathematical formulas, in particular the typesetting of mathematical symbols. These fonts are also METAFONT based. See 14.5, p. 199, for details.
5. POSTSCRIPT *Symbol Font.* This is a symbol font of the page description language POSTSCRIPT. This font is mentioned here because it contains a few mathematical symbols, but the mathematical symbols are insufficient for the typesetting of mathematical equations in TEX; see 16.10.1.2, p. 310, for details.
6. AMSFonts is a font package available from the American Mathematical Society (AMS). The AMS, a very early sponsor of TEX and METAFONT, has developed these fonts to provide additional symbols for the typesetting of mathematical equations in TEX. These fonts are available from the AMS on floppy disks in resolutions suitable for screens and several popular laser and dot-matrix printers up to 400 dots/inch; the AMS intends to make versions of 118, 180, 240, 300 and 400 dots/inch versions available to standard distributions and public archives.

   The METAFONT source code for these fonts is not included in the regular distribution, but may be requested. Some restrictions may apply.

   If you need additional information about the AMSFonts, contact the AMS at the following address: Customer Services Department, American Mathematical Society, P. O. Box 6248, Providence, RI 02940, (401) 455 4000, or send a FAX to (401) 331 3842. Email address: `cust-serv@math.ams.com`. Furthermore note that the complete distribution is available via `ftp` on node `e-maths.ams.com`. In addition to that a limited part of the distribution is available on other standard archives.

   The AMSFonts fall into three classes:

(a) *Extra math symbol fonts.* Two fonts provide extra math symbols, including uppercase Blackboard Bold.
(b) *Cyrillic fonts.* These contain the full alphabets for most slavic languages as currently rendered in cyrillic, as well as some of the pre-revolutionary letters. They are the versions newly implemented at the University of Washington, and are distributed by the AMS with permission.
(c) *Euler family of fonts (fraktur, script and cursive).* These fonts were designed by Hermann Zapf and implemented at Stanford University as part of the TEX project.

7. *Special LATEX fonts.* These are fonts special to the LATEX macro package. These fonts are used by the picture environment of LATEX and allow the user to draw a limited variety of lines and circles for simple diagrams. The user does not access these fonts directly, but only indirectly when using the picture environment.

   The names of fonts belonging to this class are `lasy5` through `lasy10`, `nlasy`, `lcircle10`, `lcirclew10`, `line10` and `linew10`. I listed those special LATEX fonts which are loaded by default (ignoring different magnifications) in LATEX (according to file `lfonts.tex` in the version of May 6, 1986). These fonts are also METAFONT-based.
8. *Miscellaneous fonts.*

   (a) There are two fonts called `black` and `gray` which are used to print proofs of characters generated by METAFONT.
   (b) Some institutions have a special *logo font* which contains only one character, the logo of their institution.

15.3.4 The Computer Modern Text Fonts

The *Computer Modern text fonts* are illustrated in Figs. 15.2–15.4, pp. 253–255. You find a sample of almost every Computer Modern text font. This should give you a rough idea of the available standard fonts of TEX. The font tables in those figures are organized as follows: the first column contains the base name of this font (the size of the font, like 10 pt, is appended to the base name leading to a complete file name of `cmr10` for Computer Modern roman at 10 pt, or to `cmr5` for the same font at 5 pt). The middle column contains the name of the font plus a short sample of text in this font.

The last column contains the "typical font sizes" available. Note the term "typical font sizes." Donald Knuth, the author of TEX and METAFONT, decided on a certain set of standard fonts. For each font, we refer to those sizes as typical sizes. Note that because the fonts are METAFONT-based, there is not a problem with generating differently sized fonts.

The figures just mentioned were produced by the following source code. This source code generates four separate pages. These pages are "electronically glued" into the text of this chapter (see Bechtolsheim (1988) for details).

$\mathcal{P}$ • **fontsam.tip** •

```
\input inputd.tip
\InputD{box-mac.tip}                        % 9.3.14, p. I-343.
\InputD{ts-dime3.tip}                       % 31.2.4.3, p. III-598.
\nonstopmode
```

No page numbers are needed.

```
\nopagenumbers
```

Now a four column table is built. The first column is not printed, but simply contains an instruction to *load* the font's `tfm` file. The second column contains the "short name" of the font, the name as it is used for `tfm`, `gf` and `pk` files. The third column (the second text column) contains the "complete name" including some sample text. The fourth and last column (the third and last column in the output) contains the typical sizes of that font.

The whole table body is stored in a macro `\FontTableComplete`, because the table must be typeset actually twice. First the sample text column is ignored so its width can be determined. Then the table is typeset.

```
\def\FontTableComplete{%
                             \tabskip = 0pt
    \halign{%
```

First column: load font.

```
        \global\font\SampleFont = ##&
```

Second column: print the font name, short version.

```
        \tt##\hfil           \tabskip = 10pt&
```

Third column: print the sample text of this font. This text should start with the "expanded" font name. For that purpose, the macro `\SampleColumn` must be defined (see below for details). `\SampleColumn` is called with the "long name" of the font as argument. This name is then printed using the sample font itself.

```
        \def\temp{##}%
        \expandafter\SampleColumn\temp&
```

Fourth and last column: typical sizes are printed here.

```
        \setbox 0 = \vtop{%
            \parindent = 0pt
            \raggedright
            \hsize = 0.75in
            ##
        }%
        \dp0 = 0pt
        \box0                \tabskip = 0pt
    \cr
```

Headings are entered now, separated from the main body of the table by a −10000 penalty for `\vsplit` to split the header later (see below).

```
        \omit&
        \omit\bf Name\hfil&
```

```
        \omit\HeadingThirdColumn&
        \omit\hfil\bf Typical Size(s)\hfil
    \cr
        \omit&
        \omit&
        \omit&
        \omit\hfil\bf (in points)\hfil
    \cr
    \noalign{\penalty -10000}
```

The header of the table ends here and the real font samples begin.

```
        cmr10& cmr& Computer Modern Roman type:1&
                5, 6, 7, 8, 9, 10, 12, 17\cr
        cmsl10& cmsl& Computer Modern Slanted Roman type:1&
                8, 9, 10, 12\cr
        cmb10& cmb& Computer Modern Boldface type:1& 10\cr
        cmbx10& cmbx& Computer Modern Boldface Extended type:1&
                5, 6, 7, 8, 9, 10, 12\cr
        cmtt10& cmtt& Computer Modern Typewriter type:1&
                8, 9, 10, 12\cr
        cmsltt10& cmsltt&
                Computer Modern Slanted Typewriter type:1& 10\cr
        cmvtt10& cmvtt&
                Computer Modern Variable Typewriter type:1& 10\cr
        cmtex10& cmtex& Computer Modern \TeX{}
                Extended ASCII type:1& 8, 9, 10\cr
        cmss10& cmss& Computer Modern Sans Serif type:1&
                8, 9, 10, 12, 17\cr
        cmssi10& cmssi& Computer Modern Slanted Sans Serif type:1&
                8, 9, 10, 12, 17\cr
        cmssdc10& cmssdc& Computer Modern Sans Serif Demibold
                Condensed type:1& 10\cr
        cmssbx10& cmssbx& Computer Modern Sans Serif
                Bold Extended type:1& 10\cr
        cmssq8& cmssq& Computer Modern Sans Serif
                Quotation Style type:1& 8\cr
        cmssqi8& cmssqi& Computer Modern Sans Serif
                Slanted Quotation Style type:1& 8\cr
        cmdunh10& cmdunh& Computer Modern Dunhill Roman type:1& 10\cr
        cmbxsl10& cmbxsl& Computer Modern
                Bold Extended Slanted Roman type:1& 10\cr
        cmff10& cmff& Computer Modern Funny Roman type:1& 10\cr
        cmfib8& cmfib& Computer Modern Fibonacci type:1& 8\cr
        cmti10& cmti& Computer Modern Text Italic type:1&
                7, 8, 9, 10, 12\cr
        cmmi10& cmmi& Computer Modern Math Italic type:1&
                5, 6, 7, 8, 9, 10, 12\cr
        cmbxti10& cmbxti& Computer Modern Bold
                Extended Text Italic type:1& 10\cr
        cmmib10& cmmib& Computer Modern Math Italic Bold type:1&
                10\cr
        cmitt10& cmitt& Computer Modern Italic Typewriter type:1&
```

```
                    10\cr
        cmu10& cmu& Computer Modern Unslanted Text Italic type:1&
                    10\cr
        cmfi10& cmfi& Computer Modern Funny Italic type:1& 10\cr
        cmcsc10& cmcsc& Computer Modern Caps and Small Caps type:1&
                    10\cr
        cmtcsc10& cmtcsc& Computer Modern Typewriter Caps and
                    Small Caps type:0& 10\cr
```

Terminate **\halign**, then **\FontTableComplete**.

```
    }%
}
```

First round of typesetting: get the width of the table, *excluding* the sample column. The heading for the sample text column must also be disabled.

```
\def\SampleColumn #1:#2{%
    \wlog{XX: Arg1 = #1}%
}
\def\HeadingThirdColumn{}
\setbox 0 = \vbox{\FontTableComplete}
```

Define a macro which contains the sample text printed in this table. If its only parameter, **#1**, is 1, a "long" version of the sample text is printed, otherwise a short version is printed.

```
\def\SampleColumnText #1{%
    :\space
    here is some sample text to display this font,
    not very long, but long enough to give you an
    idea of what output in this font looks like.%
    \ifnum #1 = 1
        \space That's it for this font.%
    \fi
}
```

For the second round of typesetting the table (with some real text in the sample text column), compute the width of that column now. Use all the available space (**\hsize** – width of all the other columns) and store it in **\dimen0**.

```
\dimen0 = \hsize
\advance\dimen0 by -\wd0
```

Define a macro for the sample text column to contain some real text. This macro has two parameters. The first one, delimited by a **:**, is the long name of the font of which a sample is printed. The undelimited parameter **#2** is **#1** of **\SampleColumnText** (the definition of **\SampleColumnText** precedes this text).

```
\def\SampleColumn #1:#2{%
    \wlog{Arg1: #1}%
    \vtop{%
        \parindent = 0pt
        \rightskip = 0pt plus 30pt
        \spaceskip = .3333em
        \xspaceskip = .5em
```

```
        \hsize = \dimen0
        \SampleFont
        \strut #1\SampleColumnText{#2}\strut
    }%
}
```

Now define a header for the third (text) column.

```
\def\HeadingThirdColumn{\hfil\bf Sample text\hfil}
```

Typeset the table and save it in a box register.

```
\setbox 0 = \vbox{\FontTableComplete}
```

Split off the table header so it can be replicated. The header is collected in box register **\HeaderBox**; see 41.3, p. IV-344, for an explanation of this process.

```
\newbox\HeaderBox
\setbox\HeaderBox = \vsplit 0 to 1000pt
\setbox\HeaderBox = \vbox{\unvbox \HeaderBox}
```

Now split the table itself into four pieces.

```
\vsize = 10in
\newdimen\SplitTableLength
\SplitTableLength = 7.0in
```

Define macro **\APage** now. It generates one sample output page. This output page is later glued into the text of this volume to become one sample font figure.

```
\def\APage{%
    \vbox{%
```

Print the header, followed by some space, followed by the font samples.

```
        \copy\HeaderBox
        \smallskip
        \vsplit 0 to \SplitTableLength
    }
    \vfill
    \eject
}
```

Print three pages of font samples.

```
\APage
\APage
\APage
\end
```

• End of `fontsam.tip` •

15.3.5 Standard Computer Modern Text Fonts, `\rm`, `\bf`, `\it`, `\tt`

`\rm`, `\bf`, `\it` and `\tt` are the most frequently used Computer Modern text fonts. The following four control sequences invoke these fonts. Note that the

Name	Sample text	Typical Size(s) (in points)
`cmr`	Computer Modern Roman type: here is some sample text to display this font, not very long, but long enough to give you an idea of what output in this font looks like. That's it for this font.	5, 6, 7, 8, 9, 10, 12, 17
`cmsl`	*Computer Modern Slanted Roman type: here is some sample text to display this font, not very long, but long enough to give you an idea of what output in this font looks like. That's it for this font.*	8, 9, 10, 12
`cmb`	**Computer Modern Boldface type: here is some sample text to display this font, not very long, but long enough to give you an idea of what output in this font looks like. That's it for this font.**	10
`cmbx`	**Computer Modern Boldface Extended type: here is some sample text to display this font, not very long, but long enough to give you an idea of what output in this font looks like. That's it for this font.**	5, 6, 7, 8, 9, 10, 12
`cmtt`	`Computer Modern Typewriter type: here is some sample text to display this font, not very long, but long enough to give you an idea of what output in this font looks like. That's it for this font.`	8, 9, 10, 12
`cmsltt`	*`Computer Modern Slanted Typewriter type: here is some sample text to display this font, not very long, but long enough to give you an idea of what output in this font looks like. That's it for this font.`*	10
`cmvtt`	Computer Modern Variable Typewriter type: here is some sample text to display this font, not very long, but long enough to give you an idea of what output in this font looks like. That's it for this font.	10
`cmtex`	`Computer Modern TeX Extended ASCII type: here is some sample text to display this font, not very long, but long enough to give you an idea of what output in this font looks like. That's it for this font.`	8, 9, 10
`cmss`	Computer Modern Sans Serif type: here is some sample text to display this font, not very long, but long enough to give you an idea of what output in this font looks like. That's it for this font.	8, 9, 10, 12, 17

Figure 15.2. Computer Modern font samples, first page.

Name	Sample text	Typical Size(s) (in points)
cmssi	*Computer Modern Slanted Sans Serif type: here is some sample text to display this font, not very long, but long enough to give you an idea of what output in this font looks like. That's it for this font.*	8, 9, 10, 12, 17
cmssdc	**Computer Modern Sans Serif Demibold Condensed type: here is some sample text to display this font, not very long, but long enough to give you an idea of what output in this font looks like. That's it for this font.**	10
cmssbx	**Computer Modern Sans Serif Bold Extended type: here is some sample text to display this font, not very long, but long enough to give you an idea of what output in this font looks like. That's it for this font.**	10
cmssq	Computer Modern Sans Serif Quotation Style type: here is some sample text to display this font, not very long, but long enough to give you an idea of what output in this font looks like. That's it for this font.	8
cmssqi	*Computer Modern Sans Serif Slanted Quotation Style type: here is some sample text to display this font, not very long, but long enough to give you an idea of what output in this font looks like. That's it for this font.*	8
cmdunh	Computer Modern Dunhill Roman type: here is some sample text to display this font, not very long, but long enough to give you an idea of what output in this font looks like. That's it for this font.	10
cmbxsl	***Computer Modern Bold Extended Slanted Roman type: here is some sample text to display this font, not very long, but long enough to give you an idea of what output in this font looks like. That's it for this font.***	10
cmff	Computer Modern Funny Roman type: here is some sample text to display this font, not very long, but long enough to give you an idea of what output in this font looks like. That's it for this font.	10
cmfib	Computer Modern Fibonacci type: here is some sample text to display this font, not very long, but long enough to give you an idea of what output in this font looks like. That's it for this font.	8

Figure 15.3. Computer Modern font samples, second page.

Name	Sample text	Typical Size(s) (in points)
cmti	*Computer Modern Text Italic type: here is some sample text to display this font, not very long, but long enough to give you an idea of what output in this font looks like. That's it for this font.*	7, 8, 9, 10, 12
cmmi	*Computer Modern Math Italic type. here is some sample text to display this font‹ not very long‹ but long enough to give you an idea of what output in this font looks like▷ Thatφs it for this font▷*	5, 6, 7, 8, 9, 10, 12
cmbxti	***Computer Modern Bold Extended Text Italic type: here is some sample text to display this font, not very long, but long enough to give you an idea of what output in this font looks like. That's it for this font.***	10
cmmib	***Computer Modern Math Italic Bold type. here is some sample text to display this font‹ not very long‹ but long enough to give you an idea of what output in this font looks like▷ Thatφs it for this font▷***	10
cmitt	*Computer Modern Italic Typewriter type: here is some sample text to display this font, not very long, but long enough to give you an idea of what output in this font looks like. That's it for this font.*	10
cmu	Computer Modern Unslanted Text Italic type: here is some sample text to display this font, not very long, but long enough to give you an idea of what output in this font looks like. That's it for this font.	10
cmfi	Computer Modern Funny Italic type: here is some sample text to display this font, not very long, but long enough to give you an idea of what output in this font looks like. That's it for this font.	10
cmcsc	COMPUTER MODERN CAPS AND SMALL CAPS TYPE: HERE IS SOME SAMPLE TEXT TO DISPLAY THIS FONT, NOT VERY LONG, BUT LONG ENOUGH TO GIVE YOU AN IDEA OF WHAT OUTPUT IN THIS FONT LOOKS LIKE. THAT'S IT FOR THIS FONT.	10
cmtcsc	COMPUTER MODERN TYPEWRITER CAPS AND SMALL CAPS TYPE: HERE IS SOME SAMPLE TEXT TO DIS- PLAY THIS FONT, NOT VERY LONG, BUT LONG ENOUGH TO GIVE YOU AN IDEA OF WHAT OUTPUT IN THIS FONT LOOKS LIKE.	10

Figure 15.4. Computer Modern font samples, third page.

instructions are really macros; see 15.5.2, p. 259, for further information. Here is a brief description of these macros.

1. `\rm` (roman font). This control sequence causes TEX to switch to the Computer Modern roman 10 pt font, the font which is normally used for the main body of a document.

   The instruction `\rm` itself is rarely given, because this font is the default and usually font changes are enclosed inside groups so there is no need for an instruction to revert back to the roman font. One case where this instruction is actually used, is discussed in 35.2.1, p. IV-91. The Computer Modern roman font is a proportionally spaced font.
2. `\it` (italic font). This control sequence switches to an italic version of the Computer Modern roman 10 pt font. It is also a proportionally spaced font. See 15.3.6 on this page for a discussion of the *italic correction.*
3. `\bf` (boldface font). This control sequences switches to a boldface version of the Computer Modern roman font. Boldface fonts are used frequently in titles and headings, and then are often used at larger than normal sizes. It is also a proportionally spaced font.
4. `\tt` (typewrite type). This control sequence invokes a typewriter-like font. It is a mono-spaced font, the only mono-spaced fonts among the four fonts listed here.

   The `\tt` font is typically used for the reproduction of computer programs and inputs, and also for the reproduction of TEX source code, as is the case in this series. This font is used under those circumstances because like the characters on a terminal screen, all characters in this font have the same width. This font therefore conveys a terminal screen-like image.

Observe that the terminology used in TEX does not exactly follow standard typesetting practice. In TEX, one says "`\bf` switches to boldface," but there is really no "boldface" font. There are only boldface *versions* of other fonts, and `\bf` invokes a boldface version of Computer Modern roman. The same remark applies to `\it`. Also note that neither `\bf\tt` nor `\tt\bf` switch to a boldface variant of the typewriter-like `\tt` font. Writing `\bf\tt`, the font change to `\bf` is immediately followed by a font change to `\tt`, thus `\bf\tt` is really the same as `\tt`.

15.3.6 Italic Correction

In an italicized font, characters are slanted to the right. When switching from an italic font back to a non-italic font (like, for instance, back to roman), the last italicized letter and the first non-italicized letter are too close together due to the slanting of the italicized letter. Here is an example to demonstrate this. The first line shows the version *without* using the italic correction ("incorrect" as far as good typesetting practices are concerned) and the second line includes the italic correction "`\/`" ("correct" as far as proper typesetting is concerned).

The difference between the two lines is admittedly quite small but nevertheless noticeable.

```
\centerline{{\it This is fun MMM} MMM and more fun}
\centerline{{\it This is fun MMM\/} MMM and more fun}
```

The following output is generated by the previous source code:

This is fun MMM MMM and more fun

This is fun MMM MMM and more fun

No italic correction is inserted when the first character outside the italic text is a period or a comma, so you would input "`{\it that}.`" and *not* "`{\it that\/}.`".

15.4 Font Sizes and Line Spacing

I will now discuss the relationship between font sizes and line spacing.

15.4.1 Font Sizes

Font sizes are normally measured in the unit point (abbreviated "pt"). There are 72.27 pt to 1 in; see 4.1.1, p. I-81, for other dimension units.

Font sizes are measured from the top of the *ascender* of a character (part of the character extending above the x-height) to the bottom of the *descender* of a character (part of the character below the baseline). No precise rules exist to determine which characters are used in these measurements. It is incorrect to simply take the maximum height and depth of all characters of a font, because sometimes the sum of the height and depth of parentheses is actually slightly larger than the font size.

Here is a little diagram which shows how the size of a font is determined (this is, by the way, one of the few places where I cheated by not using my own macros: this figure was generated using LaTeX's picture environment).

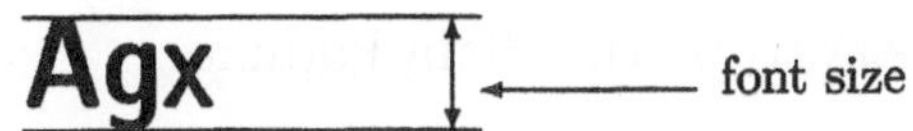

15.4.2 Typical Font Sizes

Books are typically typeset in a 10 pt font (so is this series). Technical reports sometimes use larger fonts like 11 pt or even 12 pt. Dictionaries and other reference books use smaller font sizes, to fit more text on a page. A size as small as 5 pt does not allow one to read a document comfortably for any length of time, but because reference books are not read cover-to-cover anyway this is not a real consideration.

15.4.3 Line Spacing, Leading, `\baselineskip`

The term *line spacing* defines the distance between consecutive lines in a paragraph.

Leading is the typesetter's term for specifying the line spacing, but it is *different* from the line spacing definition used above. Leading is the difference between line spacing and font size. For instance, if a manuscript is set 10/12 then the leading is 12 pt − 10 pt = 2 pt. If the leading is zero (for instance, a manuscript is set 10/10), one says that the text is set *solid*.

In TEX, `\baselineskip` usually determines the line spacing; see 7.3.4, p. I-220, for details on `\baselineskip` and other related registers controlling the line spacing. Let me now discuss settings of `\baselineskip` in relation to some fontsize f.

1. For *single spaced text* use a leading of $1.2 * f$. This is also TEX's default in the plain format: the default font in the plain format is the 10 pt Computer Modern roman font and the default value of `\baselineskip` is 12 pt. The spacing and the font size of such a manuscript would be marked as 10/12, see Chicago (1982).

   The line spacing is usually made somewhat larger than the font size, f, to avoid contact of a descender of one line with an ascender of the following line. Therefore `\baselineskip` is usually not set precisely to the font size.
2. For *one-and-a-half spaced text* use a line spacing of $1.6 * f$.
3. For *double spaced text* use a line spacing of $2.0 * f$.

15.5 Fonts in the Plain Format

Next is a discussion of how the plain format handles fonts. This discussion is the basis for a much more powerful mechanism of handling fonts, that forms the beginning of the next chapter.

15.5.1 The Fonts Loaded by the Plain Format

The following discussion is based on version 3.0 of the plain format source code. The plain format obviously uses the Computer Modern fonts. Certain point sizes in the plain format are set up to be used directly (10 pt, 7 pt and 5 pt). Others (9 pt, 8 pt and 6 pt) are pre-loaded, but to use one of those fonts the `\font` instruction for such a font has to be repeated.

For example, the plain format contains the following TEX instructions to load fonts (among others):

```
\font\tenrm   =   cmr10 % roman text
```

```
\font\preloaded  = cmr9
\font\preloaded  = cmr8
\font\sevenrm    = cmr7
\font\preloaded  = cmr6
\font\fiverm     = cmr5
```

In other words, the plain format defines control sequences `\tenrm`, `\sevenrm` and `\fiverm` to switch to a 10 pt, 7 pt or 5 pt roman font. In order to use any of the fonts defined with `\preloaded`, you have to *repeat* the `\font` instruction for these fonts. For instance, you would have to write,

```
\font\eightrm = cmr8
```

What is the advantage of this method? TEX now does *not* go out and read-in the `tfm` file of `cmr8` again. Instead, it simply uses the previously loaded, but hidden, `tfm` file.

What was just explained is a detail the user can really ignore. If you need to load a font and you are not sure whether it is loaded use `\font`. If the font was already loaded, then the `tfm` file will *not* be read-in again.

In plain TEX one can access the following fonts *without* using `\font` first:

```
\font\tenrm    = cmr10 % roman text
\font\sevenrm  = cmr7
\font\fiverm   = cmr5

\font\teni     = cmmi10 % math italic
\font\seveni   = cmmi7
\font\fivei    = cmmi5

\font\tensy    = cmsy10 % math symbols
\font\sevensy  = cmsy7
\font\fivesy   = cmsy5

\font\tenex    = cmex10 % math extension
\font\tenbf    = cmbx10 % boldface extended
\font\sevenbf  = cmbx7
\font\fivebf   = cmbx5

\font\tentt    = cmtt10 % typewriter
\font\tensl    = cmsl10 % slanted roman

\font\tenit    = cmti10 % text italic
```

Observe that none of the fonts loaded by the plain format of TEX are larger than 10 pt (which doesn't mean you can't use any fonts larger than 10 pt).

15.5.2 Font Changing Macros of the Plain Format

The plain format defines a set of font changing macros (probably you have been using these macros so far (rather than `\tenrm` and so forth) to change fonts; the instructions below are slightly modified):

```
\def\rm {\fam0 \tenrm}
\def\mit{\fam 1}
\def\oldstyle{\fam1 \teni}
\def\cal{\fam 2}
\def\it {\fam4 \tenit}
\def\sl {\fam5 \tensl}
\def\bf {\fam6 \tenbf}
\def\tt {\fam7 \tentt}
```

The important point is that the above macros do two things (take the definition of `\bf` in the following as an example):

1. Switch to the proper text font (`\tenbf`). This instruction has *no* effect in math mode.
2. Change family (`\fam6`). This instruction is only relevant in math mode and has no effect in text mode.

The `\bf` macro could be rewritten as follows perhaps making it a little easier to understand. It reads: "if math mode, change to family 6, otherwise switch to font `\tenbf`."

```
\def\bf{%
    \ifmmode
        \fam6
    \else
        \tenbf
    \fi
}
```

15.6 Summary

In this chapter we learned:

- There are two magnifications in TEX: global magnification (typically set by calling `\magnification`) and font magnification. Both are multiplied together as far as the final output is concerned. There is a standard set of magnification `\magstep 0...5`, and `\magstephalf`.
- "True dimensions" allows the user to bypass the magnification.
- There are a variety of font classes, the most important ones being the text and math fonts of the Computer Modern fonts and the POSTSCRIPT fonts.
- There are proportionally spaced fonts of which the most commonly used are Computer Modern roman (`\rm`), Computer Modern italic (`\it`), and Computer Modern Boldface (`\bf`). Character widths are different among different

characters of the same font. These fonts are normally used for right and left flush text.

- Contrary to that, characters of fixed spaced fonts (the most important one is the typewriter-like style, `\tt`) have constant width. These fonts can be used to reproduce screen images of computer output.
- Font switching instructions such as `\rm` and `\bf` are not fonts defined by `\font`. Instead they are macros which switch fonts for text and font families for mathematical equations.
- Fonts in mathematical equations are administered as font families.
- Font changes are normally enclosed inside curly braces which form a group.
- The most commonly used font size is 10 pt; single line spacing with that font is achieved by a value of 12 pt for `\baselineskip`.
- TeX can use any font for which a `tfm` file is available. The `tfm` file only contains information about the size of characters and ligature and kerning information. Fonts can be magnified.
- All Computer Modern fonts are generated by the METAFONT program which not only generates `tfm` but also `gf` files. The `gf` files are later read in by a driver to print the text.
- Characters are treated internally by TeX as numbers, based on the ASCII character code. Typically fonts in TeX have 128 characters, but 256 characters are also possible.
- Various fonts are pre-loaded in the plain format and therefore the corresponding `tfm` file need not being read in again, if the font is loaded again.

16
More on Fonts in TEX

Continuing the discussion of fonts this chapter will mainly emphasize a way of switching type faces and font sizes efficiently.

16.1 A Sophisticated Way of Organizing Fonts

Text fonts in TEX may be organized in a sophisticated manner. What will ultimately be developed is a set of macros where the specification of the *font size* and the specification of the *typeface* are separate and independent of each other. Thus there will be two separate sets of instructions, one determining the size and the other determining the typeface. A macro to load fonts on demand and a macro to define a "substitution font" will also be presented.

Note that math fonts are not included in this scheme, but they could be added easily.

16.1.1 Safe Loading of Fonts, `\NewFont`

The macro `\NewFont` which can be used like `\font`, prevents the accidental use of a control sequence name for a font name already in use somewhere else. This macro has one parameter, `#1`, the control sequence for a font which is being loaded. Otherwise, the application of this macro is identical to that of `\font`. The font specification (such as `cmr10 scaled \magstep 1`) should follow the call to this macro immediately. See 16.1.9.2, p. 272, for example applications of this macro.

$\mathcal{P}'$ • `newfont.tip` •

```
\InputD{testdef.tip}                        % 21.5.6, p. III-173.
\def\NewFont #1{%
    \if\DefinedConditional{#1}%
        \errmessage{\string\NewFont: intended font name
```

```
            "\string#1" already used.}
    \fi
    \font #1%
}
```

• End of newfont.tip •

16.1.2 Loading Fonts on Demand

Let me now discuss how fonts can be loaded *on demand*. Fonts should be loaded on demand when they are used rarely, because tfm files take up a lot of TEX's memory.

The loading of fonts on demand is done in such a way that it is *transparent* to the user: the user is *not* required to make a distinction between using a font already loaded and a font loaded on demand. Therefore, the name of the instruction to *change* to a font loaded on demand must be identical to the instruction to *load* this font. This is explained in the following Subsubsection.

16.1.2.1 The Principle of Loading Fonts on Demand

In the following example, \rnf stands for rarely needed font). Note that in the macro definition below, the macro's name (this macro will load the font) and the font's control sequence name are identical. The very first time (and the very first time only), when macro \rnf is expanded, this control sequence stands for a *macro which loads the requested font.* But \rnf does not only load the font, it also redefines \rnf for the reminder of the TEX job. Now \rnf is defined to stand for the new font.

Here is an example for a \rnf macro:

```
\def\rnf{%
    \global\font\rnf = cm...
    \rnf
}
```

The second line of the preceding macro reads \global\font... rather than \font.... The \global in this line is required because font changes typically occur inside groups, and the redefinition of \rnf must be in effect even after the group related font change is terminated. See 22.7.3, p. III-230, for some additional details.

16.1.2.2 A Macro to Load Fonts on Demand, \LoadFontOnDemand

Let me now present a macro which handles the loading of fonts on demand in more general ways. The \LoadFontOnDemand macro has two parameters:

- **#1**. The font's name, i.e., a control sequence by which a font change to the font being loaded will later be performed such as `\XIVbf`.
- **#2**. The description of the font so that TeX can load the required `tfm` file using `\font`, e.g. `cmr10 scaled \magstep 2`.

Here is the definition of this macro (see 16.1.9.2, p. 272, for details):

$\mathcal{P}'$ • `lfondem.tip` •

```
\def\LoadFontOnDemand #1#2{%
    \def #1{%
        \global\font#1 = #2\relax
        #1%
        \message{\string\LoadFontOnDemand: font \string#1
            (#2) loaded on demand.}%
    }%
}
```

• End of `lfondem.tip` •

16.1.3 Font Substitution, the Definition of `\SubstituteFontX`

Certain fonts may not be available in a desired size at a site. Of course, the user can execute METAFONT to compute such missing fonts. Here though we want to define a mechanism by which a substitution font can be defined in such a case.

The macro `\SubstituteFontX`, allows the user to declare one font to be treated in such a way that TeX will substitute another font for it. The `\SubstituteFontX` macro has the following four parameters:

- **#1**. The font's control sequence, which is being defined.
- **#2**. The name of the typeface the font belongs to (like `\sc` for small caps).
- **#3**. The size of the font (needed for an intelligent warning message).
- **#4**. The instruction to invoke the Roman font that is substituted for **#1**. Note that this macro does not allow for a different substitution.

This macro is called `\SubstituteFontX` rather than `\SubstituteFont`, because a different font substitution macro is defined further below. The macro as defined here will not work with the font setup of this series.

Here is the macro definition.

$\mathcal{P}'$ • `substf.tip` •

```
\def\SubstituteFontX #1#2#3#4{%
```

Define a macro for a font change instruction.

```
    \def#1{%
```

The very first time a font change to font **#1** occurs, a message is printed alerting the user that font substitution takes place. This message is not repeated later, however.

```
        \message{%
            \string\SubstituteFontX: No \string#2 font of
            #3pt, using \noexpand\rm instead.%
        }%
```

Here the substitution font is assigned.

```
        \global\let #1 = #4%
```

The redefinition is taken care of. Next actually invoke that font.

```
        #1%
```

End of the macro definition.

```
    }%
}
```

• End of `substf.tip` •

Let me briefly show an example application of the preceding macro: in this case the typewriter font is replaced by Computer Modern Roman 10 pt. This, of course, is not a typical application, because the typewriter font is usually available. Note that no property of the font being substituted is imported. For instance, this technique does *not* try to simulate a fixed spaced font such as `\tt` with a Roman font.

• `ex-substf.tip` •

```
\input inputd.tip
\InputD{substf.tip}                       % 16.1.3, p. 265.
\SubstituteFontX{\tt}{\tt}{10}{\rm}
This is fun, {\tt this too}. And {\tt a little more} of this.
\bye
```

• End of `ex-substf.tip` •

The preceding source code generates the following log file:

• `ex-substf.log` •

```
This is TeX, C Version 3.14 (...)
**&/usr/local/tex/lib/fmt/plain ex-substf.tip
(ex-substf.tip (inputd.tip
(namedef.tip
) (inputdl.tip
\@InputDStream=\write0
))
(substf.tip
)
\SubstituteFontX: No \tt font of 10pt, using \rm instead.
[1] )
Output written on ex-substf.dvi (1 page, 264 bytes).
```

The output generated by the preceding code reads as follows:
This is fun, this too. And a little more of this.

16.1.4 Another Font Substitution Macro

Let me introduce a *new* definition of macro `\SubstituteFont`. I will first present this new definition, and then I will explain the reason for this new definition. This macro (similar to the one defined in the preceding Subsection) has the following four parameters:

- `#1`. The font's control sequence, which is being defined.
- `#2`. The name of the typeface the font belongs to (like `\sc` for small caps).
- `#3`. The size of the font (needed for an intelligent warning message).
- `#4`. The instruction to invoke the roman font that is substituted for `#1`. Note that this macro does not allow for a different substitution.

Here is the macro definition.

$\mathcal{P}'$ • **ts-subst.tip** •

```
\InputD{substf.tip}                     % 16.1.3, p. 265.
\InputD{newoutfr.tip}                   % 4.6.2, p. I-113.
\def\SubstituteFont #1#2#3#4{%
```

We need to define a conditional, initially set to true, so that a warning message about ongoing font substitution is printed. After the first time this message will not be repeated. This conditional is initially true to indicate that the warning message should be printed.

```
    \expandafter\newifOF \csname if-\string#1\endcsname
    \csname if-\string#1true\endcsname
```

Define a macro for a font change instruction.

```
    \def#1{%
```

The very first time a font change to font `#1` occurs, a message is printed alerting the user that font substitution takes place. This message is not repeated later, however.

```
        \csname if-\string#1\endcsname
            \message{%
                \string\SubstituteFont: No \string#2 font at
                size #3pt, using \noexpand\rm instead.%
            }%
            \global\csname if-\string#1false\endcsname
        \fi
```

The substitution font is used now.

```
        #4%
    }%
}
```

• End of `ts-subst.tip` •

Let me briefly show a (rather unlikely) example application of the preceding macro: in this case the typewriter font is replaced by Computer Modern roman

10 pt. Again, as before, this substitution is unlikely in a real life example, because the \tt font is usually available at the needed sizes.

• ex-substf-2.tip •

```
\input inputd.tip
\InputD{ts-subst.tip}                    % 16.1.4, p. 267.
\SubstituteFont{\tt}{\tt}{10}{\rm}
This is fun, {\tt this too}. And {\tt a little more} of this.
\bye
```

• End of ex-substf-2.tip •

The preceding source code generates the following log file:

• ex-substf-2.log •

```
This is TeX, C Version 3.14 (...)
**&/usr/local/tex/lib/fmt/plain ex-substf-2.tip
(ex-substf-2.tip (inputd.tip
(namedef.tip
) (inputdl.tip
\@InputDStream=\write0
))
(ts-subst.tip
(substf.tip
) (newoutfr.tip
)) [1] )
Output written on ex-substf-2.dvi (1 page, 264 bytes).
```

The output generated by the preceding code reads as follows:

This is fun, this too. And a little more of this.

Note that the main difference between \SubstituteFontX and \SubstituteFont is that in the latter of the two macros no redefinition of the font macro takes place. Instead a note is made on whether this substitution was reported once before or not, and the very first time the substituted font is used a message is printed.

16.1.5 Grouping Fonts by Font Sizes

The next natural step is to discuss how fonts can be grouped into font size groups. The discussion is limited to the following five fonts: roman, italic, boldface, typewriter-like and small caps.

It is customary to name a specific font of a specific size by giving the font's size in roman numerals before the font's name itself. Capital roman numerals are used to separate the font size from the typeface's name visually.

Here is the first block of instructions defining regular 10 pt fonts:

```
\font\Xrm = cmr10
\font\Xit = cmti10
\font\Xbf = cmbx10
\font\Xtt = cmtt10
```

```
\font\Xsc = cmcsc10
```

The second block of instructions define 24 pt fonts. Observe that two fonts are loaded on demand (see 22.7.3, p. III-230), and one is not loaded at all but a substitute font is defined (see 16.1.3, p. 265). You would have to adopt the following example to your circumstances, of course.

```
\font\XXIVrm = cmr17  scaled \magstep 2
\font\XXIVbf = cmbx12 scaled \magstep 4
\LoadFontOnDemand{\XXIVit}{cmti12 scaled \magstep 4}
\LoadFontOnDemand{\XXIVtt}{cmtt12 scaled \magstep 4}
\SubstituteFont{\XXIVsc}{\sc}{24}{\XXIVrm}
```

16.1.6 Font Size Change Instructions and Typeface Change Instructions

The next step is to separate *font size changes* and *typeface changes* completely. The following font setup (used in this series) works as follows:

1. A set of macros allowing the user to switch between various font sizes is defined.
2. A separate set of macros to switch typefaces is defined. These typeface changes will change between fonts of the current size. For instance, if `\bf` is invoked and the current font size is 12 pt, then the following text is printed in a 12 pt boldface font.
3. Both types of macros are *almost* completely independent of (orthogonal to) each other.

The following additional considerations must be taken into account when the font size is changed:

1. `\baselineskip`, the line spacing, must be adjusted.
2. The current strut should be redefined and adjusted to the new value of `\baselineskip`.
3. Also provisions for a line spacing multiplication factor `\LineSpaceMultFactor` are made which defines the relationship of `\baselineskip` and the current font size.
4. Other values controlling vertical spacing such as `\abovedisplayskip` should also be adjusted (not done in the macros below).

16.1.7 Font Size Grouping Macros, `\DefineFontSizeGroup`

Let me now define a macro which is used to establish *font groups*, that is to group fonts of the same size together. This macro is called `\DefineFontSizeGroup` and has the following two parameters:

- #1. The font size as a roman numeral.
- #2. The font size in points, not necessarily an integer number, specified *without* the unit point (while #1 is assumed to be rounded to the nearest integer, the assumption is that #2 is a precise value).

The macro assumes that all the fonts you want to combine in one group are already loaded from `\font` or the previously defined macros `\NewFont`, `\LoadFontOnDemand` and `\SubstituteFont`, and that you follow the naming conventions of the preceding Section.

The macro below also assumes (as all macros in this series do) that you have already loaded the plain format. The plain format defines the math families such as `\bffam` and so forth.

$\mathcal{P}'$ • **fsized.tip** •

Save the current font size in the following dimension register.

```
\newdimen\CurrentFontSize
\InputD{namedef.tip}                         % 19.1.8, p. III-73.
```

The macro definition itself begins.

```
\def\DefineFontSizeGroup #1#2{%
```

Print an initial message.

```
     \wlog{\string\DefineFontSizeGroup: defining group "#1" (#2 pt)}%
```

Define a macro which has the following name: concatenate **FontSize** and the roman numeral font size #1. It will be explained shortly how this macro is usually applied.

```
     \NameDef{FontSize#1}{%
```

The macro `\rm` changes to family 0 and changes to the properly sized text font. This macro is obviously very similar to the definition of `\rm` of the plain format (see 15.5.2, p. 259).

```
          \def\rm{\fam = 0          \NameUse{#1rm}}%
```

Boldface, italic, typewriter-like style, small caps and slanted are handled the same way. More could be added easily.

```
          \def\bf{\fam = \bffam   \NameUse{#1bf}}%
          \def\it{\fam = \itfam   \NameUse{#1it}}%
          \def\tt{\fam = \ttfam   \NameUse{#1tt}}%
          \def\sc{\NameUse{#1sc}}%
          \def\sl{\fam = \slfam   \NameUse{#1sl}}%
```

Save the current font size in dimension register `\CurrentFontSize`. Then change `\baselineskip` to `\LineSpaceMultFactor` times the current font size.

```
          \CurrentFontSize = #2pt
          \baselineskip = \LineSpaceMultFactor\CurrentFontSize
```

Note that so far no font change has occurred. This is done by changing to the roman type at the current size.

```
            \rm
        }
}
```

By default, single line spacing is in effect. You can redefine macro `\LineSpaceMultFactor` for a different line spacing. Note that for this line spacing to become effective you need to initiate one of the font size change commands (for instance, `\normalsize`, see below), because otherwise `\baselineskip` will not be changed.

```
\def\LineSpaceMultFactor{1.2}
```

• End of `fsized.tip` •

16.1.8 Logical Font Size Names

The last step, before we present an example application of the preceding ideas, is to introduce *logical font size names* like `\small`, `\normalsize` and `\Large`. When using these instructions the user does *not* specify explicitly any font size, but uses terms like "small font" or "large font."

This approach is desirable because the user should be provided with instructions which are *independent* of the current base font size of a document. If the base size of a document is 10 pt, then a small font might be defined as an 8 pt font. If, on the other hand, the base size of a document is 12 pt, then one might choose a 10 pt font for a small font.

The last step is to introduce instructions like `\small`, `\normalsize` and others. Those instructions do nothing else but call the appropriate `\FontSizeZZ` type of macros.

16.1.9 An Example Application

For the following example we will use the logical font size names that were used in this series too.

$\mathcal{P}'$ • `ts-fonts.tip` •

```
\InputD{lfondem.tip}                    % 16.1.2.2, p. 265.
\InputD{ts-subst.tip}                   % 16.1.4, p. 267.
\InputD{fsized.tip}                     % 16.1.7, p. 270.
\InputD{newfont.tip}                    % 16.1.1, p. 263.
\InputD{testdef.tip}                    % 21.5.6, p. III-173.
```

One special kludge we need here: a font "larger than life" to print the title on the first page of the preliminaries.

```
\font\RmLargerThanLife = cmr17 scaled \magstep5
```

16.1.9.1 Math Font Family Business

First we will define math font families. The definitions taken from the plain format source code are commented out, because these font families are already set up through `plain.tex`. If you added fonts yourself, then you would do so here.

```
% \newfam\itfam          % \it is family 4
% \newfam\slfam          % \sl is family 5
% \newfam\bffam          % \bf is family 6
% \newfam\ttfam          % \tt is family 7
```

16.1.9.2 Loading Fonts

Now all the necessary fonts are loaded. Note that we use `\NewFont` and all other previously defined macros relating to font loading.

Load fonts, 5 pt first (v).

```
\NewFont\Vrm = cmr5
\NewFont\Vit = cmti7 at 5pt
\NewFont\Vbf = cmbx5
\NewFont\Vsc = cmcsc10 at 5pt
\SubstituteFont{\Vtt}{\tt}{7}{\VIIrm}
\SubstituteFont{\Vsc}{\sc}{7}{\VIIrm}
\SubstituteFont{\Vsl}{\sl}{7}{\VIIrm}
```

Load fonts, 7 pt (vii).

```
\NewFont\VIIrm = cmr7
\NewFont\VIIit = cmti7
\NewFont\VIIbf = cmbx7
\NewFont\VIIsc = cmcsc10 at 7pt
\SubstituteFont{\VIItt}{\tt}{7}{\VIIrm}
\SubstituteFont{\VIIsl}{\sl}{7}{\VIIrm}
```

Load fonts, 8 pt (viii).

```
\NewFont\VIIIrm = cmr8
\NewFont\VIIIit = cmti8
\NewFont\VIIIbf = cmbx8
\NewFont\VIIItt = cmtt8
\NewFont\VIIIsc = cmcsc10 at 8pt
\SubstituteFont{\VIIIsl}{\sl}{8}{\VIIIrm}
```

Load fonts, 9 pt (ix).

```
\NewFont\IXrm = cmr9
\NewFont\IXit = cmti9
\NewFont\IXbf = cmbx9
\NewFont\IXtt = cmtt9
\NewFont\IXsc = cmcsc10 at 9pt
\SubstituteFont{\IXsl}{\sl}{9}{\IXrm}
```

Load fonts, 10 pt (x).

```
\NewFont\Xrm = cmr10
\NewFont\Xit = cmti10
\NewFont\Xbf = cmbx10
\NewFont\Xtt = cmtt10
\NewFont\Xsc = cmcsc10
\NewFont\Xsl = cmcsc10
```

Load fonts, 12 pt (xii).

```
\NewFont\XIIrm = cmr12
\NewFont\XIIit = cmti12
\NewFont\XIIbf = cmbx12
\NewFont\XIItt = cmtt12
\NewFont\XIIsc = cmcsc10 scaled \magstep 1
\NewFont\XIIsl = cmcsc10 scaled \magstep 1
```

Load fonts, 13.1 pt (xiii).

```
\NewFont\XIIIrm = cmr12 scaled \magstephalf
\NewFont\XIIIit = cmti12 scaled \magstephalf
\NewFont\XIIIbf = cmbx12 scaled \magstephalf
\NewFont\XIIItt = cmtt12 scaled \magstephalf
\NewFont\XIIIsc = cmcsc10 scaled \magstephalf
\NewFont\XIIIsl = cmcsc10 scaled \magstephalf
```

Load fonts, 17.28 pt (xvii).

```
\NewFont\XVIIrm = cmr12 scaled \magstep 2
\NewFont\XVIIit = cmti12 scaled \magstep 2
\NewFont\XVIIbf = cmbx12 scaled \magstep 2
\NewFont\XVIItt = cmtt12 scaled \magstep 2
\NewFont\XVIIsc = cmcsc10 scaled \magstep 3
\NewFont\XVIIsl = cmcsc10 scaled \magstep 3
```

Load fonts, 20.74 pt (xxi).

```
\NewFont\XXIrm = cmr17 scaled \magstep 1
\NewFont\XXIit = cmti12 scaled \magstep 3
\NewFont\XXIbf = cmbx12 scaled \magstep 3
\NewFont\XXIsc = cmcsc10 scaled \magstep 4
\LoadFontOnDemand{\XXItt}{cmtt12 scaled \magstep 3}
\SubstituteFont{\XXIsl}{\sl}{20.74}{\XXIrm}
```

Load fonts, 24.88 pt (xxv).

```
\NewFont\XXVrm = cmr17 scaled \magstep 2
\NewFont\XXVit = cmti12 scaled \magstep 4
\NewFont\XXVbf = cmbx12 scaled \magstep 4
\NewFont\XXVsc = cmcsc10 at 5pt
\LoadFontOnDemand{\XXVtt}{cmtt12 scaled \magstep 4}
\SubstituteFont{\XXVsl}{\sl}{20.74}{\XXVrm}
```

For the preceding definitions one could have used macro names based on the use of `\csname` (see 19.1.4, p. III-69) and then used names which reflect the font size more accurately (such as "`\Group20.72`").

16.1.9.3 Define Font Size Groups

Next here is the definition of font size groups.

```
\DefineFontSizeGroup{V}{5}
\DefineFontSizeGroup{VII}{7}
\DefineFontSizeGroup{VIII}{8}
\DefineFontSizeGroup{IX}{9}
\DefineFontSizeGroup{X}{10}
\DefineFontSizeGroup{XII}{12}
\DefineFontSizeGroup{XIII}{13.14}
\DefineFontSizeGroup{XVII}{17.28}
\DefineFontSizeGroup{XXI}{20.74}
\DefineFontSizeGroup{XXV}{24.88}
```

16.1.9.4 Define Logical Font Size Groups

Define the logical font size names `\tiny`, `\scriptsize` and so forth. These font definitions refer to a 10 pt basic font size, and naturally would have to be done differently if a different base font size were chosen.

```
\let\tiny = \FontSizeV
\let\scriptsize = \FontSizeVII
\let\footnotesize = \FontSizeVIII
\let\small = \FontSizeIX
\let\normalsize = \FontSizeX
\let\large = \FontSizeXII
\let\Large = \FontSizeXIII
\let\LARGE = \FontSizeXVII
\let\huge  = \FontSizeXXI
\let\Huge  = \FontSizeXXV
```

16.1.9.5 "Closing Arguments"

Set-up single line spacing now. Remember that the value of `\baselineskip` is determined by the product of the current font size and `\LineSpaceMultFactor`.

```
\def\LineSpaceMultFactor{1.2}
\normalsize
```

• End of `ts-fonts.tip` •

Besides the problem that no scaled math fonts are included in the preceding setup note also that in a format like LATEX font size definitions such as `\let\tiny = \FontSizeV` are stored in style files so that different styles (with different base sizes) can be defined.

16.1.10 Extension and Discussion of the Discussed Font Organization

I would like to make two remarks at the end of this Section. First math fonts have not been included in the above set-up, something which could be changed quite easily. Second of all is the currently common practice of using magnified simply wrong. Magnification is intended for magnifying a font so that upon photographic reduction; see 15.2.11, p. 241, for details. What really should happen is that one uses METAFONT at all times to generate fonts at the desired sizes for whatever resolution(s) one needs.

16.2 Interword Spacing

The issue of interword space is not only of interest in the discussion of fonts, but also important to the discussion of the typesetting of paragraphs.

16.2.1 Spaces in the Text Are Translated into Interword Glue

Recall that when the text of a paragraph is processed by TeX, TeX reduces multiple adjacent spaces to single spaces (more precisely single space tokens), which are then translated into *interword glue*. Therefore, as far as TeX is concerned, the space between words is filled by *interword glue*, not by a space character.

TeX's concept of glue is a mathematical description of a distance which can "breathe," stretch and shrink; see 5.1, p. I-121, for details on glue. Whether the interword glue can actually stretch or shrink depends on the currently active font. Obviously, if a text is set left flush and right flush, then the interword space must be allowed to stretch and shrink.

Let us look at an example where this adjustment of the interline glue is clearly visible. The width of the paragraph to be typeset is small. Therefore, the interword space needs to stretch and shrink a lot (typesetting narrow columns is discussed in more detail later in 16.2.4, p. 277). Here is the source code of the example (the various settings of parameters will be explained later):

```
$$\vbox{
    \parindent = 0pt              \hsize = 2.0in
    \hyphenpenalty = 10000        % Prevents hyphenation.
    \spaceskip = 4pt plus 20pt minus 2pt
    \xspaceskip = 1.5\spaceskip
    This is a sample text to explain the concept of interword glue.
    \TeX{} thinks of the whole paragraph as one big line.
    This line is broken up into smaller lines each 2.0~in wide.
    Compare, in particular, the space between the words of the first
```

```
    and second line.
}
$$
```

This source code generates the following text:

> This is a sample text to explain the concept of interword glue. TEX thinks of the whole paragraph as one big line. This line is broken up into smaller lines each 2.0 in wide. Compare, in particular, the space between the words of the first and second line.

Compare the first and the second line in the preceding output; you can see that there is a large difference in the amount of space between the words of those two lines.

16.2.2 Typesetting a Paragraph

How TEX processes a paragraph can be summarized as follows (details can be found in 10.1.2, p. 3; what follows is a gross oversimplification). This explanation is provided to illustrate the importance and effect of interword glue.

1. TEX accumulates the whole paragraph as *one long line.* Each space was replaced by interword glue while the paragraph was read-in.
2. This line is broken into lines of approximately the same length (`\hsize`).
3. The line length of each line is set to the desired value (`\hsize`). The interword glue is stretched out, if a line is too short, and is pushed together, if the line is too long.

16.2.3 A Summary of The Rules About Interword Space Computation

The interword space computation of TEX can be summarized as follows. Note that this list does *not* explain how the space factor f was computed in the first place; see 16.2.7, p. 279, for details.

1. TEX keeps track of a positive integer called the *space factor* f. This factor is similar to the global magnification factor `\mag` in that the factor represents a value 1000 times the amount of multiplication. This factor is usually 1000 and is recomputed as each character in a word is processed according to the rules explained below.

2. Let's first discuss the case where $f < 2000$. In this case, the amount of interword space is determined as follows:

   (a) Use the value from parameter `\spaceskip`, if its value is not zero. Multiply the shrinkability of this glue by $1000/f$ and the stretchability by $f/1000$.

   The value of `\spaceskip` is *usually* zero and therefore this computation does usually *not* apply.

   (b) Otherwise, if `\spaceskip` is zero, use the values from the current font, thus using a glue determined by font dimension parameters 2, 3 and 4 (see 15.2.15, p. 243). Then modify the shrinkabilities and stretchabilities according to the rules just established for `\spaceskip`.

3. If $f \geq 2000$, the preceding explanation for $f < 2000$ still applies with two minor modifications:

   (a) It is now the parameter `\xspaceskip` that is tested for whether it contains a zero glue. The default value for this register is also zero.

   If this glue is non-zero, note that the stretchability or shrinkability of the glue specified in it is *not* scaled (*not* multiplied by $1000/f$ or $f/1000$).

   (b) If `\xspaceskip` is zero, then, for the computation of the interword glue, the "extra space" (font dimension parameter 7) is added to the length of the regular interword space. Other than that nothing changes, the multiplications with $1000/f$ and $f/1000$ are performed.

The stretchabilities and shrinkabilities of the interword glues of Computer Modern font which are proportionally spaced (`\tenrm`, `\tenit` and `\tenbf`), are rather small, as you can see from the table of 15.2.15, p. 243. Therefore the interword glue can stretch and shrink only by small amounts before TeX complains about underfull and overfull hboxes. These tight values encourage TeX to typeset well. If TeX cannot find good line breaks within the given tolerances of the interword glue, then TeX will report this rather than trying to fudge it in some way. Some people think that these tolerances are too tight, but good typesetting requires that the interword glue shrink and stretch by small amounts only. Good typesetting also requires sometimes the rewriting of a sentence or two.

16.2.4 An Example of User-Defined `\spaceskip`

Let me now present an example in which text in a rather narrow column is typeset left-flush and right-flush. `\spaceskip` is set to a value which allows the interword glue to stretch and shrink a lot ("a lot" compared to the default interword glue of an ordinary TeX font). This is one way of coping with narrow columns.

Here is the source code of the example:

```
$$\vbox{
```

```
        \hsize = 1.65in     \spaceskip = 3.333pt plus 20pt
        \xspaceskip = 1.5\spaceskip
        This text is typeset in a narrow column.
        The interword space is allowed to breathe a lot.
        And this should actually be enough for an example.
    }
$$
```

This source code generates the following output:

This text is typeset in a narrow column. The interword space is allowed to breathe a lot. And this should actually be enough for an example.

16.2.5 French Spacing, Extended Space, `\frenchspacing`, `\nonfrenchspacing`

The "end of sentence space" is the space inserted between the end of a sentence character, like a period, and the next word. This space can be treated in two different ways:

1. It can be treated as a *regular space*. This is called *French spacing*. In order to have TEX use this spacing invoke the macro **`\frenchspacing`**. This type of spacing is used in this series.
2. If you want to have *extended spacing*, where that space is made slightly larger than the regular interword space, call the macro **`\nonfrenchspacing`**. This type of spacing is TEX's default.

For the following discussion assume that "non-French spacing" is in effect. Note that *typing* two spaces after a symbol ending a sentence in the input will *not* tell TEX to insert an extended space, because multiple adjacent spaces are reduced to single spaces (more accurately space tokens). It is still desirable to end a sentence with two spaces, because it improves the readability of your input.

16.2.6 Another Look at `\spaceskip`

In TEX's Computer Modern proportionally spaced fonts, the extended interword space is usually 33% larger than the regular interword space. In case of the typewriter-like `\tt` font, the extended space is *twice* the size of the regular interword space. In this font the space after a period ending a sentence is therefore twice the regular interword space. This setup obviously implements the common typewriter practice of leaving two spaces after a period ending a sentence.

There are cases where one does *not* want the extra space to be inserted between a period and the following text. This is the case for initials: X. Y. Miller. This issue is discussed in the next Subsection, item 2 on this page.

16.2.7 Space Factor Computation, `\spacefactor`, `\sfcode`

Above we explained how, given a certain space factor, the space between words is computed. The details of the space factor computation itself follow.

Note that in TeX *each character* has a space factor code. This code is an integer and is assigned by the `\sfcode` primitive. For instance, `\sfcode‘A = 999` assigns space factor 999 to the character "A."

As some text is processed, there is a *current space factor, f*, an integer, which is determined as follows:

- The initial value of f is 1000.
- Usually the space factor f is taken from the *space factor* of the last character (let's call it g). The following are exceptions to this rule.
  - If $g = 0$, then f is *not* changed.
  - If $f < 1000 < g$, then f is set to 1000. This means that the space factor never changes from a value below 1000 to a value above 1000 in one step but requires at least two steps.

The next question to address, the effect of the space factor on the interword glue, was discussed in 16.2.3, p. 276.

Let me now discuss the space factors of the individual characters, how they are set up by default and then slightly modified by the plain format.

1. The default for all characters is 1000 with the exception of all *uppercase* letters "`A`" through "`Z`" (only) which have a default of 999.
2. The space factor of the period is 3000, which usually means that the space after a period is extended. However, because of the space factors of uppercase letters in a text which reads "`A. B. Miller`," no extended space is inserted after the initials in this name. Note while this is usually the desired effect, this can lead to errors if some text reads as follow: "**`Now look at example A. This is a very good example`**. In this case, the extended space must be manually programmed (see below how this can be done).
3. The space factor of the closing parenthesis is set to zero by the plain format, which means that the space factor of the character preceding the closing parenthesis is used. So in a sentence like "**`some words.)`**" the extended space resulting from the period is not influenced by the closing parenthesis and therefore an extended space is printed after the closing parenthesis.
4. For other examples see the source code of the plain format.

You can change the value of the space factor by assigning the desired value to `\spacefactor`. You can also print out the current value, as the following example shows. `\spacefactor` can be referred to only in horizontal mode and not inside a write, so we need to copy its value into a counter register before writing it to the log file.

• `ex-spacef.tip` •

```
\def\WS #1{%
    #1%
    {%
        \count0 = \spacefactor\relax
        \wlog{%
            \string\WS (#1): spacefactor = \the\count0
                \space now.%
        }%
    }%
}
```

Here are the sample calls to the preceding macro.

```
\nonfrenchspacing
\WS{A}\WS{.}\WS{B}\WS{.}\WS{ }%
\WS{W}\WS{i}\WS{l}\WS{l}\WS{i}\WS{a}\WS{m}\WS{.}%
\WS{G}\WS{o}\WS{o}\WS{d}\WS{ }%
\WS{n}\WS{i}\WS{g}\WS{h}\WS{t}\WS{.}\WS{ }%
\bye
```

• End of `ex-spacef.tip` •

This code generates the following log file.

• `ex-spacef.log` •

```
This is TeX, C Version 3.14 (...)
**&/usr/local/tex/lib/fmt/plain ex-spacef.tip
(ex-spacef.tip
\WS(A): spacefactor = 999 now.
\WS(.): spacefactor = 1000 now.
\WS(B): spacefactor = 999 now.
\WS(.): spacefactor = 1000 now.
\WS( ): spacefactor = 1000 now.
\WS(W): spacefactor = 999 now.
\WS(i): spacefactor = 1000 now.
\WS(l): spacefactor = 1000 now.
\WS(l): spacefactor = 1000 now.
\WS(i): spacefactor = 1000 now.
\WS(a): spacefactor = 1000 now.
\WS(m): spacefactor = 1000 now.
\WS(.): spacefactor = 3000 now.
\WS(G): spacefactor = 999 now.
\WS(o): spacefactor = 1000 now.
\WS(o): spacefactor = 1000 now.
\WS(d): spacefactor = 1000 now.
\WS( ): spacefactor = 1000 now.
\WS(n): spacefactor = 1000 now.
\WS(i): spacefactor = 1000 now.
```

```
\WS(g): spacefactor = 1000 now.
\WS(h): spacefactor = 1000 now.
\WS(t): spacefactor = 1000 now.
\WS(.): spacefactor = 3000 now.
\WS( ): spacefactor = 3000 now.
[1] )
Output written on ex-spacef.dvi (1 page, 236 bytes).
```

16.2.8 The Definitions of Macros `\frenchspacing` and `\nonfrenchspacing`

Now that the space factor computation has been explained, I present the definitions of the `\frenchspacing` and `\nonfrenchspacing` macros of the plain format.

```
\def\frenchspacing{%
    \sfcode'\. = 1000
    \sfcode'\? = 1000
    \sfcode'\! = 1000
    \sfcode'\: = 1000
    \sfcode'\; = 1000
    \sfcode'\, = 1000
}
\def\nonfrenchspacing{%
    \sfcode'\. = 3000
    \sfcode'\? = 3000
    \sfcode'\! = 3000
    \sfcode'\: = 2000
    \sfcode'\; = 1500
    \sfcode'\, = 1250
}
```

It should be quite obvious how the `\frenchspacing` macro achieves French spacing: characters which can appear at the end of a sentence are assigned a space factor of 1000, and therefore loose their special "extended spacing property."

16.2.9 The Implicit Globalness of Space Factor Computations, `\SaveSpaceFactor`, `\RestoreSpaceFactor`

Note that the space factor computation, as just discussed, is *implicitly global* in that the space factor assignment does not follow grouping (this makes perfect sense because the space factor computation should not be influenced by groups due to font changes). There are cases where the space factor needs to be saved and later restored (because the save and restore effect cannot be achieved with a group). For that purpose, we define the two macros, `\SaveSpaceFactor` and `\RestoreSpaceFactor`. The first one saves the space factor, the second one restores it (the definition of `\footnote` in the plain format uses the same technique).

$\mathcal{P}'$ • `spacefac.tip` •

```
\catcode`\@ = 11
\def\SaveSpaceFactor{%
```

`\xdef` is used to achieve a global effect and also to force the expansion of `\the\spacefactor` when `\SaveSpaceFactor` is called.

```
    \xdef\@SavedSpaceFactor{%
        \spacefactor = \the\spacefactor
    }%
}
\def\RestoreSpaceFactor{%
    \@SavedSpaceFactor
}
\catcode`\@ = 12
```

• End of `spacefac.tip` •

16.2.10 Another Look at Extended Spaces and Periods

Note that because of the choice of spacing factors in a text like `A. B. Miller` no extended space will be inserted.

Next look at the following examples:

1. `Abc def`: regular space between the two words.
2. `Abc\space def`: same, because `\space` expands to a space token, like the regular space did in the preceding example.
3. `Xyz. xxx`. Extended space is inserted.
4. `Xyz.\space xxx`. Also in this case an extended space is inserted.
5. `Xyz.\ xxx`. In this case a *regular* space is inserted.
6. `Xyz.~xxx`. In this case also a regular space is inserted, mainly because except for a penalty inserted by `~` (and not by `\ `), this and the preceding example are identical.

16.2.11 "Characters Per Pica"

To be able to estimate how much space a text will take one uses "characters per pica" in typesetting. Here is a macro `\CharactersPerPica` which has one argument, `#1`, a font and which prints out for this font the number of characters per pica as a decimal number (with two positions after the decimal point).

$\mathcal{P}'$ • font-cpp.tip •

```
\InputD{leadingz.tip}                     % 3.3.8.1, p. I-50.
\InputD{imodn.tip}                        % 3.3.13, p. I-58.
\def\CharactersPerPica #1{%
```

The following text has 423 characters. This count *includes* spaces and end-of-lines, but excludes the leading white spaces in every line.

```
    \setbox0 = \hbox{%
        #1\relax
        This paragraph has 423 characters. We know that
        because we counted it. It is very simple to count, because
        the Emacs editor we are using has a function ``advance
        by one character.'' And with the prefix command (that's
        Emacs terminology) you can execute ``advance by
        one character'' 423~times. Now, let's hope that this text
        is representative of ordinary text so that our average
        number of characters per pica is correct.
    }%
    {%
```

We need to perform the following computation. If w_{sp} is the width of the text in scaled points, and w the width in picas, then we know

$$w = \frac{w_{sp}}{65536 * 12}$$

We furthermore know that the number of characters per pica ch is computed as follows:

$$ch = \frac{423}{w} = 423 * \frac{65536 * 12}{w_{sp}}$$

We furthermore multiply the preceding number by 100 (TEX has no floating point numbers) and before printing the number divide the number by 100 and use "mod 100" to compute the two decimal digits after the decimal point. The final formula to compute therefore is

$$\frac{423 * 65536 * 12 * 100}{w_{sp}}$$

The preceding formula would lead to overflow computing the numerator in a 32 bit machine and therefore has to be rewritten as follows:

$$\frac{423 * 65536 * 12}{\frac{w_{sp}}{100}}$$

`\count1` contains the width of the text in scaled points. Divide it by 100.

```
        \count1 = \wd0
        \divide\count1 by 100
```

Save the number of characters in `\count0`. Multiply with 12 and 65536.

```
        \count0 = 423
```

```
            \multiply\count0 by 65536
            \multiply\count0 by 12
```

The division leaving in count0 the number of characters per pica times 100.

```
            \divide\count0 by \count1
```

Now compute the part before the fractional point in \count2 and the part after the fractional part in \count0.

```
            \count2 = \count0
            \divide\count2 by 100
            \IModN{\count0}{100}{\count3}%
```

Print the result.

```
            \the\count2.\LeadingZ{\count3}%
        }%
}
```

• End of font-cpp.tip •

Here is an example application of the preceding macro.

• ex-font-cpp.tip •

```
\input inputd.tip
\InputD{font-cpp.tip}                    % 16.2.11, p. 283.
\unskip\space
font \Verb+\tenrm+ has on the average \CharactersPerPica{\tenrm}
characters per pica.
\bye
```

• End of ex-font-cpp.tip •

The preceding source code generates the following output:

font \tenrm has on the average 2.73 characters per pica.

16.3 Printing Fonts Used in TeX Documents

Next we discuss how a characters of certain fonts in a dvi file are actually printed on a piece of paper or displayed on a workstation screen.

16.3.1 Raster Output Devices

Recall that TeX's typesetting is solely based on the *dimensions* of characters. The *shape* of characters is not known to TeX. The shapes of characters are also *not* contained in dvi files. The dvi file only contains instructions on *where* to print the characters of a document and what fonts to use for this purpose.

Practically all output devices used in the context of TEX are *raster output devices*. Their working principle is the main focus in this Subsection. A raster output device consists of a large rectangular grid of *pixels*. Each pixel can be individually controlled by the output device and it can be printed either black or white. There is no way to print only half a pixel or to assign a gray value to a pixel: a pixel is either on or off, black or white.

The distance between two adjacent pixels is the main factor to determine the quality of the output. This value is referred to as the *resolution* of the output device, and it is normally given as the number of pixels per inch.

Fig. 16.1 on the next page shows the pixel matrix of a raster output device with the lower part of a "Z" embedded into it. From this figure you can see that is it only possible to *approximate* non-horizontal and non-vertical lines.

A raster device blackens or whitens its pixels. In the upper left corner of Fig. 16.1 on the next page you can see how a small black rectangle is composed of single black pixels. In real life, these pixels overlap and are not precisely square (in fact they are more round than square). In this figure pixels are drawn slightly smaller, so that the effect of filling an area by filling-in adjacent pixels is clearly visible.

In Fig. 16.2, p. 287, you can see the lower part of a "Z" with the pixels filled in. The pixel matrix of a raster output device *digitally approximates* the letter "Z". We human beings (almost) do not perceive this as an "approximation": the human eyes are not sharp enough to discriminate among the individual pixels. An eagle, a bird famous for his good eyes, would probably complain and tell you to print the document on an output device of higher resolution if he were to look at such a document closely.

There are three classes of output raster devices that are of interest in the context of TEX. They are listed in the order of increasing resolution, and therefore, in the order of increasing output quality:

1. *Raster-scan displays.* Such raster scan displays are found at computer work stations and PCs. A typical resolution for 19 in screens is 114 dots/inch (812 x 1024 pixels).
2. *Laser printers.* They are by far the most frequently used output device for TEX output; their resolution is, in almost all cases, 300 dots/inch, although laser printers with higher resolutions are available.

   One quick side remark: if you would actually measure the resolution of such a device you would be surprised to discover that the resolution may be off by a percent or two. This is because of inaccuracies in the optical system of laser printers.
3. *Photo typesetters.* These output devices have varying resolution with values up to 4500 dots/inch, normally starting at 1200 dots/inch.

   The reason for the high quality of photo typesetter output is not only its high resolution, but also the fact that the output is printed on film rather than paper.

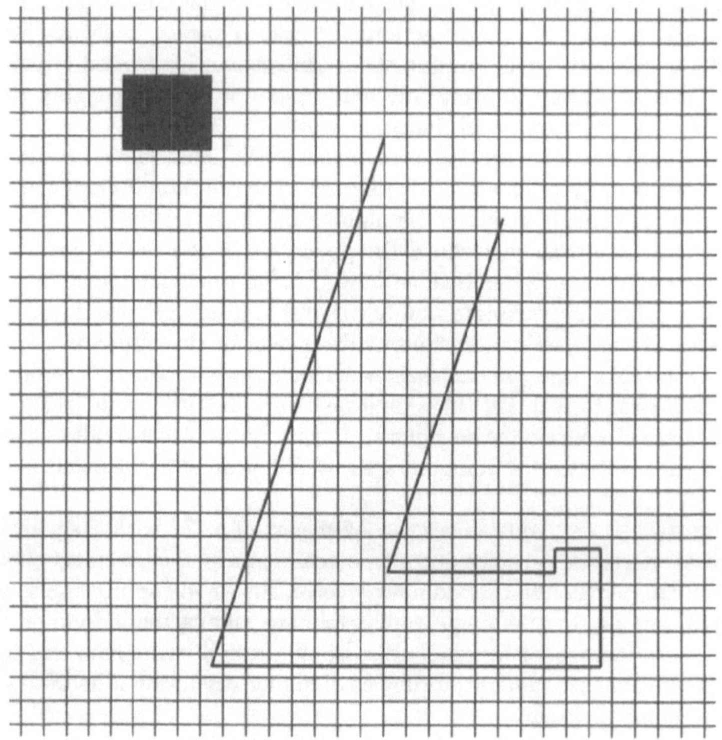

Figure 16.1. Lower part of a "Z" embedded into the pixel grid of a laser printer.

16.3.2 Source of Pixel of Fonts in TEX Documents

There are three ways to have access to the pixel information of fonts used in TEX.

1. The pixel data is stored in so-called *pixel files.* The pixel data must be sent to the printer each time a document is printed. It is the driver which is responsible for initiating the downloading of pixel patterns.

   This setup of printing a TEX document is by far the most frequent one; 16.3.3 on the next page deals with it in more detail.
2. The pixel data is stored in *font cartridges*, which are plugged into the printer. The HP LaserJet is a laser printer with this property.
3. The characters are stored in an *outline format,* normally inside the printer or on a disk attached to the printer which can be accessed by the printer directly. From this outline format the pixel patterns for each character are computed. One advantage of this approach is that it allows this outline to

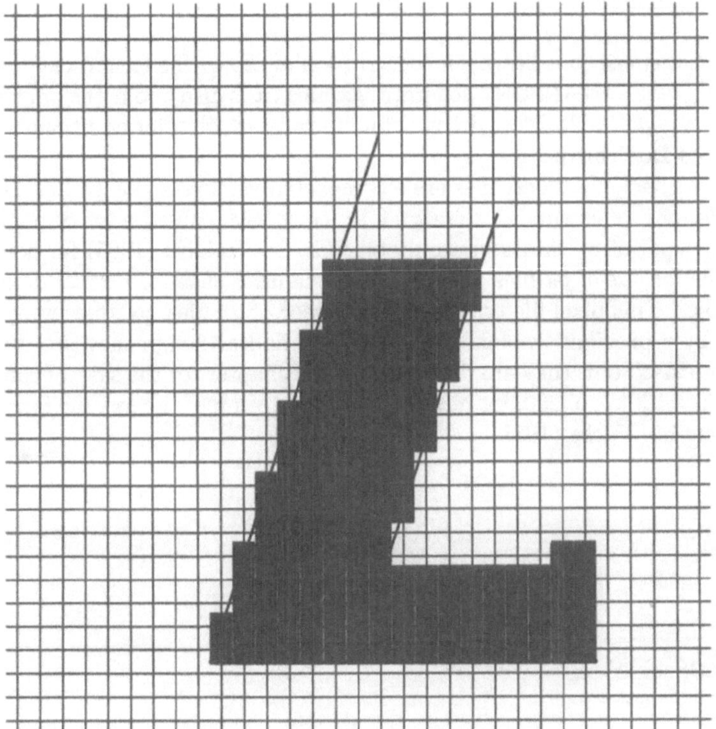

Figure 16.2. Lower part of a "Z" with pixels filled in.

be used for the same font at almost any size, that is the font can be scaled to any size. POSTSCRIPT printers use this approach. See 16.10.1, p. 309, for details.

To be truthful, there is a fourth way of storing the shapes of characters, on optical disks (traditional phototypesetters use this approach), but we will not discuss it any further here.

16.3.3 Pixel Files of METAFONT-Based Fonts

If pixel patterns from pixel files are used, then these patterns must be transmitted to the printer used for output. This process typically happens anew with each document. Within in each document however, the pixel patterns for a particular character are usually downloaded only once because they are saved in the printer's memory.

In TEX there are three different types of pixel files:

1. **gf** files (generic font files). The description of the **gf** file format can be found in GF (1985). METAFONT generates **gf** files (it is also about the only way **gf** files are generated). Most drivers can interpret **gf** files; for additional details, see 16.3.4 on this page.
2. **pk** files (packed). These files are a compacted version of the **gf** files. Because **pk** files are much smaller than **gf** files, they have gained a lot of popularity on all types of machines, not only on PCs. See Rokicki (1985) for details on the file format of **pk** files and 16.3.6 on the next page for further details.
3. **pxl** files. This pixel file type is a predecessor of **gf** files. **pxl** files were generated by a previous version of METAFONT, and although they are supposed to be phased out, they are discussed in this chapter for the sake of completeness; see 16.3.7, p. 290, for details. The description of the **pxl** file format can be found in Fuchs (1981).

16.3.4 GF Files

From what has been discussed so far, you know that a font, let's say **cmr10** (Computer Modern roman 10 point) has precisely one **tfm** file. The same font has a variety of **gf** files. There are three reasons why fonts can and usually do have multiple **gf** files:

1. Different *resolutions* of output devices require different pixel patterns, and therefore different **gf** files.
2. Depending on the printer engine of an output device, different **gf** files may be needed. There are write-white and write-black engines and those engines require different **gf** (or **pk** and **pxl**) files.
3. If a particular font is used at different *magnifications*, then for each of those magnifications, a different pixel file is needed.

You will need to distinguish among these different **gf** files of one and the same font. To deal with items 1. and 2. of the preceding list, one normally stores their respective **gf** files in separate directories. For 3., a different method is used. One has to be able to distinguish among different **gf** files of the same font, but for different magnifications. This is done by an (integer) number, called the *numerical file extension* in this series, as part of the file extension of the **gf** file names. The basename of the **gf** and the **tfm** file name are, of course, identical. For instance, for **cmr10** you might have **gf** files **cmr10.300gf** and **cmr10.329gf**. The next Subsection shows how this numerical file extension is computed.

16.3.5 The Numerical File Extension

The numerical file extension of **gf** and **pk** files are discussed the same way, but it is different from the computation of the numerical extension of `pxl` files.

First assume a document is not magnified (the global magnification is 1.0) and the font for which we want to compute the numerical file extension is also not magnified. Then, the numerical file extension of the **gf** file to be used for this font is the *resolution* of the printer used for the output. For instance, to print a text in Computer Modern roman 10 pt on a laser printer with a resolution of 300 dots/inch (no global magnification and no font magnification), the **gf** file to be used is `cmr10.300gf`. In case the laser printer had a resolution of 600 dots/inch, the driver would load the pixel patterns of this font from **gf** file `cmr10.600gf`.

Let me now present the formula which specifies how the numerical file extension of a **gf** file is computed in general. Assume the global magnification of a document is *mag* and the magnification of the font for which this computation is performed is *scal*. *res* is the resolution of the output device. Note that *mag* and *scal* are specified in "TEX's scale," i.e., for a magnification of 1.0 you would use a value of 1000 in the formula. *ngf* stands for the numerical file extension of **gf** (and **pk**) files.

$$\begin{aligned} ngf(mag, scal, res) &= \text{round}\left(res * \frac{mag}{1000} * \frac{scal}{1000}\right) \\ &= \text{round}\left(\frac{res * mag * scal}{1,000,000}\right) \end{aligned}$$

For further details about the file type of **gf** files see sections 1142–1148 of Knuth (1986d). This part of Knuth (1986d) is also reprinted in GF (1985). Typical numerical file extension values can be found in the last column (ngf) of Table 16.1 on the next page.

16.3.6 PK Files

Pk files are essentially equivalent to **gf** files. The most important difference is a different encoding of the pixel pattern information which in **pk** files is based on run length encoding. This leads to a considerable reduction (typically in the neighborhood of 50%) in the size of **pk** files compared to **gf** files.

pk files are generated from **gf** files by the **gftopk** utility, *not* by METAFONT directly. The amount of additional computation required to decode **pk** files in the driver is negligible and accepted as a necessary tradeoff in comparison with the saved disk space. Because of these savings in disk space, **pk** files are very popular.

The numerical file extension of **pk** files and **gf** files are identical. See the previous Subsection on **gf** files for details. See Rokicki (1985) for a detailed description of the file format of **pk** files.

Table 16.1. Magnification and pixel file extension table for 300 dots/inch printers.

`\magstep...`	Ratio	*TEX′smag*	Increase (%)	*npxl*	*ngf*
`\magstep 0`	$1.2^0 = 1.0$	1000	0	1500	300
`\magstephalf`	$1.2^{0.5} = 1.095$	1095	≈ 10	1642	329
`\magstep 1`	$1.2^1 = 1.2$	1200	≈ 20	1800	360
`\magstep 2`	$1.2^2 = 1.44$	1440	≈ 50	2160	432
`\magstep 3`	$1.2^3 = 1.728$	1728	≈ 75	2592	518
`\magstep 4`	$1.2^4 = 2.073$	2073	≈ 100	3111	622
`\magstep 5`	$1.2^5 = 2.488$	2488	≈ 150	3732	746

16.3.7 PXL Files

Pxl files can be described as an old version of **gf** files. Because **gf** files are more general (for instance, in **gf** files one can have an almost unlimited number of characters, whereas **pxl** files allow for 128 characters per font only), **pxl** files are on their way out. Nevertheless here is the formula for the computation of the numerical file extension *npxl* for **pxl** files (see Fuchs (1981) for more details). The numerical file extension of **pxl** files is the *magnification factor* of a document, if the resolution of the printing device were *200 dots/inch.* Let us assume a document with no global or font magnification using font **amr10**. The driver used to print the document would try to access **cmr10.1000pxl**, if the printer has a resolution of 200 dots/inch. If the global magnification of the document would be 2000 (2.0), the driver would access **amr10.2000pxl** instead.

Here is the formula for the computation of the numerical file extension of **pxl** files (*mag*, *scal* and *res* are defined in equation 16.3.5 on the previous page).

$$npxl(mag, scal, res) = \text{round}\left(\frac{res}{200} * \frac{mag}{1000} * \frac{scal}{1000} * 1000\right)$$
$$= \text{round}\left(\frac{res * mag * scal}{200000}\right)$$

From that follows that if there is no global magnification and no font magnification in effect, then the numerical file extension of **pxl** files is computed as follows:

$$PXL'(res) = PXL(res, 1000, 1000) = res * 5$$

In Table 16.1 on this page you can find an overview of which **pxl** files are needed with a standard laser printer (having a resolution of 300 dots/inch).

16.4 Ligatures, Kerning

Ligatures and kerning are discussed next. They are discussed together because they are related to each other.

16.4.1 Ligatures

A *ligature* is a pair (or occasionally also a triple) of characters like "fi," which when typeset, are treated as *one* character. Ligatures are *not* multiple characters printed closely adjacent to other, although the designer of a font, when drafting a ligature, probably started out: drawing the two or three characters of a ligature close together. Which character combinations form ligatures is font-dependent. Typical ligatures besides "ffi" are "fi," "ff," "fl" and "ffl."

TeX has no built-in set of ligatures because ligatures are totally font dependent. It is the font designer which decides which characters form ligatures and which not, and the ligature information is therefore stored in `tfm` files.

Monospaced fonts like the typewriter-like font (`\tt`) usually do not have ligatures, because having ligatures contradicts the idea of mono-spacing, and ordinary typewriters do not have ligatures. So while, in the regular Computer Modern roman font, words like "final" and "difficult" contain ligatures, this is not the case using the typewriter font of TeX printing the same words: "`final`" and "`difficult`."

16.4.2 Kerning

Kerning, in typesetting, means a small change in the spacing between individual characters. Observe that the kerning and the ligature concepts are different: ligatures are the combination of two or three characters into one character; in kerning, the spacing between two characters (that are treated as two separate characters) is changed.

The amount of horizontal space inserted between two characters varies among different character pairs. In most cases, the spacing is decreased, thereby printing the two characters closer together than where two bounding boxes would normally suggest.

The standard example for applying kerning is the combination of the two characters "A" and "V," which are typically moved closer together by kerning. With kerning (by simply entering "`AVA`", TeX does the kerning automatically), the output reads "AVA"[1], while without kerning the output looks as follows:

---

[1] I actually cheated a little bit by inserting some additional negative horizontal space to move the characters even closer than they would have been printed through

"AVA."

The kerning information is contained in the **tfm** file of a font. It is therefore font dependent, and the amount of kerning is under control of the designer of a font. TEX has *no* notion of "A" and "V" being in general a pair of characters where kerning is applied in general.

Applying kerning only makes sense in proportionally spaced fonts, since monospaced fonts do not use kerning because kerning contradicts the idea of monospacing.

When TEX inserts kern between two characters, then **\kern** is inserted, not horizontal glue; see 5.4.8, p. I-140, for details.

To find out about the kerning tables of a particular font, use **tftopl**, a utility that converts the **tfm** file of a font into its property list. This property list contains, among other information, the kerning tables of a font.

16.4.3 Suppressing Ligatures and Kerning

TEX can be instructed to insert not ligatures or kerning (which is, with the exception of a series like this, of little interest). However it can be done by inserting an empty group "**{}**" between the characters involved in the ligature or kerning process. These characters are then no longer adjacent to each other as far as TEX is concerned. For instance, to print the "unkerned" version of "AVA" enter "**A{}V{}A**." The other possibility to suppress kerning and the forming of ligatures is to enclose each character inside a separate hbox:

```
        \hbox{\hbox{A}\hbox{V}\hbox{A}}
```

16.4.4 A Ligature and Kerning Related Example

Here is a brief example showing kerning and ligatures in a different way. Some TEX input is saved inside an hbox and then the content of this hbox is subsequently written to the log file using **\showbox**. The input,

• **showkern.tip** •

```
\input inputd.tip
\InputD{shboxes.tip}                    % 4.5.15, p. I-111.
\setbox 0 = \hbox{AVA A{}V{}A fi f{}i ffi f{}f{i} \tt ffi}
\ShowBoxDepthOne{0}
\bye
```

• End of **showkern.tip** •

generates the following log file (observe the **\kern** instructions and the ligature information):

---

kerning.

• showkern.log •

```
This is TeX, C Version 3.14 (...)
**&/usr/local/tex/lib/fmt/plain showkern.tip
(showkern.tip (inputd.tip
(namedef.tip
) (inputdl.tip
\@InputDStream=\write0
))
(shboxes.tip
)
> \box0=
\hbox(6.94444+0.0)x107.13889
.\tenrm A
.\kern-1.11113
.\tenrm V
.\kern-1.11113
.\tenrm A
.\glue 3.33333 plus 1.66498 minus 1.11221
.\tenrm A
.\tenrm V
.\tenrm A
.\glue 3.33333 plus 1.66498 minus 1.11221
.\tenrm ^^L (ligature fi)
.\glue 3.33333 plus 1.66666 minus 1.11111
.\tenrm f
.\tenrm i
.\glue 3.33333 plus 1.66666 minus 1.11111
.\tenrm ^^N (ligature ffi)
.\glue 3.33333 plus 1.66666 minus 1.11111
.\tenrm f
.\tenrm f
.\tenrm i
.\glue 3.33333 plus 1.66666 minus 1.11111
.\tentt f
.\tentt f
.\tentt i

<to be read again>
}
l.4 \ShowBoxDepthOne{0}

)
No pages of output.
```

16.5 Accents in Text

TEX makes the distinction between *text accents* in regular text such as inside a paragraph or hbox and *math accents*. Text accents are the focus of the following discussion; see 13.4.1, p. 177, for details on math accents.

16.5.1 `\accent` to Generate an Accent

To produce a text accent, use the `\accent` control sequence. The instruction format of this primitive is as follows:

`\accent` ⟨number⟩ ⟨character⟩,

where ⟨number⟩ is the character code of the character forming the accent. The accent will appear above ⟨character⟩. When TEX places an accent, it takes into account the slanting of the character over which the accent is placed.

Here is an example: `\accent21 A` prints Ă.

Note that the instruction format given above does not tell you the full story. Here are the complete details:

1. Between `\accent` ⟨number⟩ and ⟨character⟩ you are allowed to insert other tokens, in particular a *font change*. This allows you to take the accent and the character being accented from two different fonts. For instance, to place a boldface "˘" accent on top of an "A" in the `\tt` font, enter the following: `{\bf \accent21 \tt A}` that prints Ă.
2. If the instruction right after `\accent` is a begin group instruction (like an opening curly brace), then TEX assumes that the accent is supposed to be printed "by itself," as an ordinary character. For instance, `\accent21{}` prints ˘.

Let me briefly discuss the dimensions of the accented character which is generated using `\accent`. The *width* of the accent character is disregarded as far as the width of the whole construct is concerned. After the accented character was printed TEX moves to the right by the width of ⟨character⟩. The *height* of the accented character is the sum of the height of the character being accented and the height of the accent itself.

There are, by the way, no direct provisions for "underaccents." This expression "underaccent" does not really exist, on the other hand you probably understand what I mean.

16.5.2 The Accent Definitions of the Plain Format

So the user does not have to resort to using numerical codes for accents, the plain format defines some accent macros. These macros are listed in Table 16.2 on the next page. This table also contains examples of accent applications.

The text accents used in Table 16.2 on the next page are defined as follows.

```
\def\'#1{{\accent18 #1}}
```

Table 16.2. Text accent table.

Here is the standard set of text accents in TeX.

`\`o`	ò	Grave accent
`\'o`	ó	Acute accent
`\^o`	ô	Circumflex or "hat"
`\"o`	ö	Umlaut or dieresis
`\~`	õ	Tilde or "squiggle"
`\.o`	ȯ	Dot accent
`\u o`	ŏ	Brave accent
`\H o`	ő	Long Hungarian umlaut
`\t oo`	o͡o	Tie after accent

```
\def\'#1{{\accent19 #1}}
\def\v#1{{\accent20 #1}} \let\^^_=\v
\def\u#1{{\accent21 #1}} \let\^^S=\u
\def\=#1{{\accent22 #1}}
\def\^#1{{\accent94 #1}} \let\^^D=\^
\def\.#1{{\accent95 #1}}
\def\H#1{{\accent"7D #1}}
\def\~#1{{\accent"7E #1}}
\def\"#1{{\accent"7F #1}}
\def\t#1{{\edef\next{\the\font}\the\textfont1\accent"7F\next#1}}
```

16.5.3 Dotless Characters

TeX's Computer Modern text fonts contain a dotless i and j which are needed for accented i's and j's. They are generated by `\i` (ı) and `\j` (ȷ) respectively.

16.6 Character Codes

TeX's handling of characters are based on the ASCII code as discussed in 3.3.2, p. I-40. Internally, characters are treated as numbers in the range 0...255. To print a character, you normally just type in the character, and TeX converts the character into a number based on the ASCII code. The fact that some of the characters are handled in a special way, like, for instance, the backslash which is TeX's escape character, has to do with the *category code* of this particular character. For details on category codes see 18.1.6, p. III-5.

16.6.1 \char

Under certain circumstances (for instance, to print font tables listing all characters of a font systematically) it would be desirable to print a character based on its character code, instead of providing the character itself. The `\char` control sequence allows you to do just that. `\char` ⟨number⟩ prints the character, whose code is ⟨number⟩. See 3.3.1, p. I-39, for entering such a number in hexadecimal or octal format.

Also, if one accesses one of the more exotic fonts, `\char` comes in very handy. For instance, to print two musical symbols ♭, ♯ of font `cmmi10` you would have to enter `\char'133, \char'135`. By the way, `cmmi10` is a math font and not a regular text font. By using `\char` there was no problem accessing single characters in it.

16.9, p. 302, contains numerous font tables from which you can find the character codes of many of the standard characters. The default category codes for most of the characters are listed there too.

16.6.2 Computing the Character Code

There is a way of computing the character code of a character without looking into any font tables: `` `\A``, for instance, produces the character code of the "A" wherever a numerical constant is expected. Here is a short example to print the numerical code of "A": the input `` {\count0=`\A{}\the\count0} `` prints 65. This is also discussed in 3.3.1, item 3, p. I-40. For a discussion of the necessity of inserting an empty group (empty pair of curly braces) see 3.3.3.1, p. I-44.

16.6.3 TEX Fonts Usually Have 128 Characters

Generally TEX's fonts, that is the Computer Modern fonts, have 128 characters. You are by no means restricted to 128 characters, but you can have fonts of 256 characters, in particular after the introduction of TEX 3.0. One important example of fonts with 256 characters are the PostScript fonts (16.10.1, p. 309).

To access characters with character code in the range of 128–255 (and naturally also all other characters) you have two choices:

1. Use the `\char` primitive; see 16.6.1 on this page for details.
2. Use the hexadecimal notation as described in 3.3.2.4, p. I-42, thus writing `^^xy`, where `x` and `y` are two hexadecimal digits (use lower case if not digits).

16.6.4 Unavailable Characters, `\tracinglostchars`

It is possible for fonts to have character codes for which no character is defined. This does *not* apply to the Computer Modern fonts, if character codes in the range of 0–127 are used, because the Computer Modern fonts have characters defined for each character code in the range 0–127.

If the user requests a character in a font, for which this font has no character defined, then TeX does two things:

1. The missing character is reported in the log file if this option is enabled and `\tracinglostchars` is set to a positive value (this is the default of the plain format). If this counter parameter is zero or negative the error is never reported.
2. No output occurs to the `dvi` file and no horizontal movement in the text of the current paragraph takes place.

16.7 Underlining

There are some serious limitations in TeX as far as the underlining of some text is concerned. Let me summarize briefly:

1. There is no `\underl` type of instruction, which could be applied in a font-like fashion, for instance, as follows: `Text {\underl more text} text` and "more text" would be underlined.
2. The `\underbar` macro (6.10.1, p. I-200) includes the text to underline inside an hbox. Therefore the text cannot be broken across lines. This macro is of little practical value when underlining text in a paragraph.
3. There are two other solutions to this problem:
   (a) Use a `dvi` file processor (see Bechtolsheim (1989)).
   (b) Add the underline to each character in the font.

16.8 Standard Font Table Macros

The TeXbook presents macros for the printing of font tables which I would like to discuss now. I have made some changes to these macros. In addition to renaming them I have added the facility of printing font tables for fonts with 256 characters.

16.8.1 Macro Source Code

$\mathcal{P}'$ • `fotable.tip` •

```
\catcode`\@ = 11
```

Some auxiliary macros to print a number as an octal or hexadecimal number (using the "TEX format" with a preceding single or double quote) are defined first.

```
\def\@OctPrintFontTable#1{%
    \hbox{%
        \rm\'{}%
        \kern-.2em
        \it #1\/%
        \kern.05em
    }%
}
\def\@HexPrintFontTable#1{%
    \hbox{\rm\H{}\tt#1}%
}
```

Next some auxiliary macros are defined.

```
\def\@OddLineFontTable#1{%
    \cr
    \noalign{\nointerlineskip}
    \multispan{19}\hrulefill&
    \setbox0 = \hbox{%
        \lower 2.3pt\hbox{%
            \@HexPrintFontTable{#1x}%
        }%
    }%
    \smash{\box0}%
    \cr
    \noalign{\nointerlineskip}
}
\def\@EvenLineFontTable{\cr\noalign{\hrule}}
\def\@FontTableStrut{\lower4.5pt\vbox to 14pt{}}
```

The following macro `\BeginFontTable` has one parameter, `#1`, a font change instruction. This macro is called to begin a font table of font `#1`.

```
\def\BeginFontTable #1{%
    $$
    \postdisplaypenalty = 0
    \global\count@=0
    #1
    \halign to\hsize\bgroup
        \@FontTableStrut##\relax    \tabskip = 0pt plus 10pt&
        &\hfil##\hfil&\vrule##%
    \cr
    \lower6.5pt\null
    &&&
```

```
    \@OctPrintFontTable0&&
    \@OctPrintFontTable1&&
    \@OctPrintFontTable2&&
    \@OctPrintFontTable3&&
    \@OctPrintFontTable4&&
    \@OctPrintFontTable5&&
    \@OctPrintFontTable6&&
    \@OctPrintFontTable7&
    \@EvenLineFontTable
}
```

After the `\BeginFontTable` macro one should call `\NormalFontTable`, followed by either `\EndFontTable` (for fonts with 128 characters) or `\MoreFontTable`, then `\EndFontTable` (for fonts with 256 characters).

```
\def\EndFontTable{%
    \raise 11.5pt\null
    &&&
    \@HexPrintFontTable 8&&
    \@HexPrintFontTable 9&&
    \@HexPrintFontTable A&&
    \@HexPrintFontTable B&&
    \@HexPrintFontTable C&&
    \@HexPrintFontTable D&&
    \@HexPrintFontTable E&&
    \@HexPrintFontTable F&
    \cr
    \egroup
    $$%
}
```

Again some auxiliary macros.

```
\def\:{%
    \setbox0 = \hbox{%
        \char\count@
    }%
    \ifdim\ht0 > 7.5pt
        \@RepositionFontTable
    \else
        \ifdim\dp0 > 2.5pt
            \@RepositionFontTable
        \fi
    \fi
    \box0
    \global\advance\count@ by 1
}
\def\@RepositionFontTable{%
    \setbox0 = \hbox{%
        $
            \vcenter{%
                \kern 2pt
                \box0
                \kern 2pt
```

```
            }
          $%
    }%
}
```

The macro `\NormalFontTable` (no arguments) prints the font table for character codes in the range '0–'177 (see above for an explanation of when this macro is called).

```
\def\NormalFontTable{%
    &\@OctPrintFontTable{00x}&&\:&&\:&&\:&&\:&&\:&&\:&&
        \:&&\:&&\@OddLineFontTable0
    &\@OctPrintFontTable{01x}&&\:&&\:&&\:&&\:&&\:&&\:&&
        \:&&\:&\@EvenLineFontTable
    &\@OctPrintFontTable{02x}&&\:&&\:&&\:&&\:&&\:&&\:&&
        \:&&\:&&\@OddLineFontTable1
    &\@OctPrintFontTable{03x}&&\:&&\:&&\:&&\:&&\:&&\:&&
        \:&&\:&\@EvenLineFontTable
    &\@OctPrintFontTable{04x}&&\:&&\:&&\:&&\:&&\:&&\:&&
        \:&&\:&&\@OddLineFontTable2
    &\@OctPrintFontTable{05x}&&\:&&\:&&\:&&\:&&\:&&\:&&
        \:&&\:&\@EvenLineFontTable
    &\@OctPrintFontTable{06x}&&\:&&\:&&\:&&\:&&\:&&\:&&
        \:&&\:&&\@OddLineFontTable3
    &\@OctPrintFontTable{07x}&&\:&&\:&&\:&&\:&&\:&&\:&&
        \:&&\:&\@EvenLineFontTable
    &\@OctPrintFontTable{10x}&&\:&&\:&&\:&&\:&&\:&&\:&&
        \:&&\:&&\@OddLineFontTable4
    &\@OctPrintFontTable{11x}&&\:&&\:&&\:&&\:&&\:&&\:&&
        \:&&\:&\@EvenLineFontTable
    &\@OctPrintFontTable{12x}&&\:&&\:&&\:&&\:&&\:&&\:&&
        \:&&\:&&\@OddLineFontTable5
    &\@OctPrintFontTable{13x}&&\:&&\:&&\:&&\:&&\:&&\:&&
        \:&&\:&\@EvenLineFontTable
    &\@OctPrintFontTable{14x}&&\:&&\:&&\:&&\:&&\:&&\:&&
        \:&&\:&&\@OddLineFontTable6
    &\@OctPrintFontTable{15x}&&\:&&\:&&\:&&\:&&\:&&\:&&
        \:&&\:&\@EvenLineFontTable
    &\@OctPrintFontTable{16x}&&\:&&\:&&\:&&\:&&\:&&\:&&
        \:&&\:&&\@OddLineFontTable7
    &\@OctPrintFontTable{17x}&&\:&&\:&&\:&&\:&&\:&&\:&&
        \:&&\:&\@EvenLineFontTable
}
```

The macro `\MoreFontTable` prints the part of the font table that covers character codes in the range of '200–'377 (this is necessary for PostScript fonts, but not for the regular Computer Modern fonts).

```
\def\MoreFontTable{%
    &\@OctPrintFontTable{20x}&&\:&&\:&&\:&&\:&&\:&&\:&&
        \:&&\:&&\@OddLineFontTable8
    &\@OctPrintFontTable{21x}&&\:&&\:&&\:&&\:&&\:&&\:&&
        \:&&\:&\@EvenLineFontTable
```

```
    &\@OctPrintFontTable{22x}&&\:&&\:&&\:&&\:&&\:&&\:&&
        \:&&\:&&\@OddLineFontTable9
    &\@OctPrintFontTable{23x}&&\:&&\:&&\:&&\:&&\:&&\:&&
        \:&&\:&\@EvenLineFontTable
    &\@OctPrintFontTable{24x}&&\:&&\:&&\:&&\:&&\:&&\:&&
        \:&&\:&&\@OddLineFontTable A
    &\@OctPrintFontTable{25x}&&\:&&\:&&\:&&\:&&\:&&\:&&
        \:&&\:&\@EvenLineFontTable
    &\@OctPrintFontTable{26x}&&\:&&\:&&\:&&\:&&\:&&\:&&
        \:&&\:&&\@OddLineFontTable B
    &\@OctPrintFontTable{27x}&&\:&&\:&&\:&&\:&&\:&&\:&&
        \:&&\:&\@EvenLineFontTable
    &\@OctPrintFontTable{30x}&&\:&&\:&&\:&&\:&&\:&&\:&&
        \:&&\:&&\@OddLineFontTable C
    &\@OctPrintFontTable{31x}&&\:&&\:&&\:&&\:&&\:&&\:&&
        \:&&\:&\@EvenLineFontTable
    &\@OctPrintFontTable{32x}&&\:&&\:&&\:&&\:&&\:&&\:&&
        \:&&\:&&\@OddLineFontTable D
    &\@OctPrintFontTable{33x}&&\:&&\:&&\:&&\:&&\:&&\:&&
        \:&&\:&\@EvenLineFontTable
    &\@OctPrintFontTable{34x}&&\:&&\:&&\:&&\:&&\:&&\:&&
        \:&&\:&&\@OddLineFontTable E
    &\@OctPrintFontTable{35x}&&\:&&\:&&\:&&\:&&\:&&\:&&
        \:&&\:&\@EvenLineFontTable
    &\@OctPrintFontTable{36x}&&\:&&\:&&\:&&\:&&\:&&\:&&
        \:&&\:&&\@OddLineFontTable F
    &\@OctPrintFontTable{37x}&&\:&&\:&&\:&&\:&&\:&&\:&&
        \:&&\:&\@EvenLineFontTable
}
\catcode'\@ = 12
```

• End of fotable.tip •

16.8.2 Font Table Print Example

The following source code was used to to generate the font table in Fig. 16.3 on the next page.

```
\input inputd.tip
\InputD{fotable.tip}                    % 16.8.1, p. 298.
\BeginFontTable{\it}
\NormalFontTable
\EndFontTable
```

	'0	'1	'2	'3	'4	'5	'6	'7	
'00x	Γ	Δ	Θ	Λ	Ξ	Π	Σ	Υ	"0x
'01x	Φ	Ψ	Ω	ff	fi	fl	ffi	ffl	
'02x	ı	ȷ	`	´	ˇ	˘	¯	˚	"1x
'03x	¸	ß	æ	œ	ø	Æ	Œ	Ø	
'04x	-	!	"	#	£	%	&	'	"2x
'05x	(	)	*	+	,	-	.	/	
'06x	0	1	2	3	4	5	6	7	"3x
'07x	8	9	:	;	¡	=	¿	?	
'10x	@	A	B	C	D	E	F	G	"4x
'11x	H	I	J	K	L	M	N	O	
'12x	P	Q	R	S	T	U	V	W	"5x
'13x	X	Y	Z	[	“	]	^	˙	
'14x	‘	a	b	c	d	e	f	g	"6x
'15x	h	i	j	k	l	m	n	o	
'16x	p	q	r	s	t	u	v	w	"7x
'17x	x	y	z	–	—	˝	~	¨	
	"8	"9	"A	"B	"C	"D	"E	"F	

Figure 16.3. A font table example based on the macros presented in this series.

16.9 "Comparative" Font Tables

All characters can be printed using `\char`, as was discussed before. So, to print Φ, we simply enter any of the three instructions `\char"8` (hexadecimal code), `\char'10` (octal) or `\char8` (decimal). All the possibilities are shown in the font tables on pp. 304–307. The fourth row in each of the character rows shows an alternative way of generating the character codes *directly*, in other words, without resorting to the `\char` control sequence. For instance, instead of `\char"8`, you can enter `^^(`.

16.9.1 A Brief Explanation of the Font Tables

Let us now discuss the combined font tables for the following fonts: `cmr10` (`\rm`), `cmbx10` (`\bf`), `cmit10` (`\it`) and `cmtt10` (`\tt`). The first three fonts are listed together, whereas `\tt` is listed separately, because the first three fonts are proportionally spaced fonts printing similar characters in all cases, whereas the typewriter-like font is a monospaced font leading to different output. For instance, for character code '14, the "fi" ligature is printed for the three proportionally spaced fonts, but "↓" is printed when the typewriter-like font is used.

The font tables can be found in Figs. 16.4–16.7, pp. the next–307. The tables also list the category codes for each character; category codes are discussed in 18.1.6, p. III-5.

16.9.2 Macros Used to Print the Font Tables

Let me now reprint the TEX source code to print the font tables just mentioned. You should be able to modify the following source code to print your own font tables (this source code does not consists of only macro definitions that would have to be loaded using **\InputD** rather the following code is a separate TEX program). The figures just mentioned were generated in a separate TEX run and then overlaid with a program, which allows the merging of dvi files; see 33.3.1, p. IV-42, for details.

P • **fonttab.tip** •

```
\input inputd.tip
\InputD{hex.tip}                              % 27.1.4.1, p. III-404.
\InputD{oct.tip}                              % 27.1.4.3, p. III-406.
\InputD{setstrut.tip}                         % 7.4.3.1, p. I-239.
\InputD{verb-bas.tip}                         % 18.3.1, p. III-27.
\nopagenumbers
```

Set-up parameters and the strut used in the following font tables.

```
\baselineskip = 13pt
\ComputeStrut
\def\VruleS{\MyStrut width 0.5pt}
```

The macro **\PrintCatCode** prints the category code and the character of character code **#1**. Note that the parameter is a character *code* and *not* a character. The macro deals with the "**@**" separately, because this character can have either category code 11 or 12; see 18.1.12, p. III-13.

```
\def\PrintCatCode #1{%
    \ifnum '\@ = #1\relax
        11 / 12\relax
    \else
        \the\catcode #1\relax
    \fi
}
```

Two temporary counter registers are needed for the following macro.

```
\newcount\tcount
\newcount\bcount
```

\OneChar is a macro that prints the table entry for one character. It has one parameter, **#1**, the character code of the character entry to be printed.

Hex	"0	"1	"2	"3	"4	"5	"6	"7
Octal	'0	'1	'2	'3	'4	'5	'6	'7
Decimal	0_{10}	1_{10}	2_{10}	3_{10}	4_{10}	5_{10}	6_{10}	7_{10}
\catcode	9	8	12	12	12	12	12	12
\rm, ...	Γ Γ Γ	Δ Δ Δ	Θ Θ Θ	Λ Λ Λ	Ξ Ξ Ξ	Π Π Π	Σ Σ Σ	Υ Υ Υ
\tt	Γ	Δ	Θ	Λ	Ξ	Π	Σ	Υ

Hex	"8	"9	"A	"B	"C	"D	"E	"F
Octal	'10	'11	'12	'13	'14	'15	'16	'17
Decimal	8_{10}	9_{10}	10_{10}	11_{10}	12_{10}	13_{10}	14_{10}	15_{10}
\catcode	12	10	12	7	13	5	12	12
\rm, ...	Φ Φ Φ	Ψ Ψ Ψ	Ω Ω Ω	ff ff ff	fi fi fi	fl fl fl	ffi ffi ffi	ffl ffl ffl
\tt	Φ	Ψ	Ω	↑	↓	'	¡	¿

Hex	"10	"11	"12	"13	"14	"15	"16	"17
Octal	'20	'21	'22	'23	'24	'25	'26	'27
Decimal	16_{10}	17_{10}	18_{10}	19_{10}	20_{10}	21_{10}	22_{10}	23_{10}
\catcode	12	12	12	12	12	12	12	12
\rm, ...	ı ı ı	ȷ ȷ ȷ	` ` `	´ ´ ´	ˇ ˇ ˇ	˘ ˘ ˘	¯ ¯ ¯	˚ ˚ ˚
\tt	ı	ȷ	`	´	ˇ	˘	¯	˚

Hex	"18	"19	"1A	"1B	"1C	"1D	"1E	"1F
Octal	'30	'31	'32	'33	'34	'35	'36	'37
Decimal	24_{10}	25_{10}	26_{10}	27_{10}	28_{10}	29_{10}	30_{10}	31_{10}
\catcode	12	12	12	12	12	12	12	12
\rm, ...	¸ ¸ ¸	ß ß ß	æ æ æ	œ œ œ	ø ø ø	Æ Æ Æ	Œ Œ Œ	Ø Ø Ø
\tt	¸	ß	æ	œ	ø	Æ	Œ	Ø

Figure 16.4. Font table, character codes '0–'37.

Hex	"20	"21	"22	"23	"24	"25	"26	"27
Octal	’40	’41	’42	’43	’44	’45	’46	’47
Decimal	32_{10}	33_{10}	34_{10}	35_{10}	36_{10}	37_{10}	38_{10}	39_{10}
\catcode	10	12	12	6	3	14	4	12
\rm, ...	- - -	! ! !	” ” ”	# # #	$ $ £	% % %	& & &	’ ’ ’
\tt	␣	!	"	#	$	%	&	’

Hex	"28	"29	"2A	"2B	"2C	"2D	"2E	"2F
Octal	’50	’51	’52	’53	’54	’55	’56	’57
Decimal	40_{10}	41_{10}	42_{10}	43_{10}	44_{10}	45_{10}	46_{10}	47_{10}
\catcode	12	12	12	12	12	12	12	12
\rm, ...	(((	)))	* * *	+ + +	, , ,	- - -	. . .	/ / /
\tt	(	)	*	+	,	-	.	/

Hex	"30	"31	"32	"33	"34	"35	"36	"37
Octal	’60	’61	’62	’63	’64	’65	’66	’67
Decimal	48_{10}	49_{10}	50_{10}	51_{10}	52_{10}	53_{10}	54_{10}	55_{10}
\catcode	12	12	12	12	12	12	12	12
\rm, ...	0 0 0	1 1 1	2 2 2	3 3 3	4 4 4	5 5 5	6 6 6	7 7 7
\tt	0	1	2	3	4	5	6	7

Hex	"38	"39	"3A	"3B	"3C	"3D	"3E	"3F
Octal	’70	’71	’72	’73	’74	’75	’76	’77
Decimal	56_{10}	57_{10}	58_{10}	59_{10}	60_{10}	61_{10}	62_{10}	63_{10}
\catcode	12	12	12	12	12	12	12	12
\rm, ...	8 8 8	9 9 9	: : :	; ; ;	¡ ¡ ¡	= = =	¿ ¿ ¿	? ? ?
\tt	8	9	:	;	<	=	>	?

Figure 16.5. Font table, character codes ’40–’77.

Hex	"40	"41	"42	"43	"44	"45	"46	"47
Octal	'100	'101	'102	'103	'104	'105	'106	'107
Decimal	64_{10}	65_{10}	66_{10}	67_{10}	68_{10}	69_{10}	70_{10}	71_{10}
\catcode	11 / 12	11	11	11	11	11	11	11
\rm, ...	@ **@** *@*	A **A** *A*	B **B** *B*	C **C** *C*	D **D** *D*	E **E** *E*	F **F** *F*	G **G** *G*
\tt	@	A	B	C	D	E	F	G

Hex	"48	"49	"4A	"4B	"4C	"4D	"4E	"4F
Octal	'110	'111	'112	'113	'114	'115	'116	'117
Decimal	72_{10}	73_{10}	74_{10}	75_{10}	76_{10}	77_{10}	78_{10}	79_{10}
\catcode	11	11	11	11	11	11	11	11
\rm, ...	H **H** *H*	I **I** *I*	J **J** *J*	K **K** *K*	L **L** *L*	M **M** *M*	N **N** *N*	O **O** *O*
\tt	H	I	J	K	L	M	N	O

Hex	"50	"51	"52	"53	"54	"55	"56	"57
Octal	'120	'121	'122	'123	'124	'125	'126	'127
Decimal	80_{10}	81_{10}	82_{10}	83_{10}	84_{10}	85_{10}	86_{10}	87_{10}
\catcode	11	11	11	11	11	11	11	11
\rm, ...	P **P** *P*	Q **Q** *Q*	R **R** *R*	S **S** *S*	T **T** *T*	U **U** *U*	V **V** *V*	W **W** *W*
\tt	P	Q	R	S	T	U	V	W

Hex	"58	"59	"5A	"5B	"5C	"5D	"5E	"5F
Octal	'130	'131	'132	'133	'134	'135	'136	'137
Decimal	88_{10}	89_{10}	90_{10}	91_{10}	92_{10}	93_{10}	94_{10}	95_{10}
\catcode	11	11	11	12	0	12	7	8
\rm, ...	X **X** *X*	Y **Y** *Y*	Z **Z** *Z*	[**[** /	“ **“** *“*	] **]** /	^ **^** *^*	˙ **˙** *˙*
\tt	X	Y	Z	[	\	]	^	_

Figure 16.6. Font table, character codes '100–'137.

Hex	"60	"61	"62	"63	"64	"65	"66	"67
Octal	'140	'141	'142	'143	'144	'145	'146	'147
Decimal	96_{10}	97_{10}	98_{10}	99_{10}	100_{10}	101_{10}	102_{10}	103_{10}
\catcode	12	11	11	11	11	11	11	11
\rm, ...	‘ ‘ ‘	a **a** *a*	b **b** *b*	c **c** *c*	d **d** *d*	e **e** *e*	f **f** *f*	g **g** *g*
\tt	`	a	b	c	d	e	f	g

Hex	"68	"69	"6A	"6B	"6C	"6D	"6E	"6F
Octal	'150	'151	'152	'153	'154	'155	'156	'157
Decimal	104_{10}	105_{10}	106_{10}	107_{10}	108_{10}	109_{10}	110_{10}	111_{10}
\catcode	11	11	11	11	11	11	11	11
\rm, ...	h **h** *h*	i **i** *i*	j **j** *j*	k **k** *k*	l **l** *l*	m **m** *m*	n **n** *n*	o **o** *o*
\tt	h	i	j	k	l	m	n	o

Hex	"70	"71	"72	"73	"74	"75	"76	"77
Octal	'160	'161	'162	'163	'164	'165	'166	'167
Decimal	112_{10}	113_{10}	114_{10}	115_{10}	116_{10}	117_{10}	118_{10}	119_{10}
\catcode	11	11	11	11	11	11	11	11
\rm, ...	p **p** *p*	q **q** *q*	r **r** *r*	s **s** *s*	t **t** *t*	u **u** *u*	v **v** *v*	w **w** *w*
\tt	p	q	r	s	t	u	v	w

Hex	"78	"79	"7A	"7B	"7C	"7D	"7E	"7F
Octal	'170	'171	'172	'173	'174	'175	'176	'177
Decimal	120_{10}	121_{10}	122_{10}	123_{10}	124_{10}	125_{10}	126_{10}	127_{10}
\catcode	11	11	11	1	12	2	13	15
\rm, ...	x **x** *x*	y **y** *y*	z **z** *z*	– – –	— — —	˝ ˝ ˝	˜ ˜ ˜	¨ ¨ ¨
\tt	x	y	z	{	|	}	~	¨

Figure 16.7. Font table, character codes '140–'177.

```
\def\OneChar #1{%
    \tcount = #1
    \advance\tcount by \bcount
    \vtop{%
```

The following `\hsize` specification determines the width of one character entry in the table. It was found by trial and error so that the table would actually fit onto the current page.

```
        \hsize = 0.54in
        \offinterlineskip
        \hrule
        \line{\VruleS\hfil \tt"\Hex{\tcount}\hskip0.3\hsize\VruleS}
        \hrule
        \line{\VruleS\hfil \tt'\Oct{\tcount}\hfil\VruleS}
        \hrule
        \line{\VruleS\hfil $\the\tcount_{\scriptscriptstyle 10}$%
                                                  \hfil\VruleS}
        \hrule
        \line{\VruleS\hfil \PrintCatCode{\tcount}\hfil\VruleS}
        \hrule
        \line{\VruleS                 \hfil
                         {\rm \char\tcount}\hskip 2.3pt
                         {\bf \char\tcount}\hskip 2.3pt
                         {\it \char\tcount}%
                         \hfil\VruleS}
        \hrule
        \line{\VruleS     \hfil\tt \char\tcount\hfil\VruleS}
        \hrule
    }%
}
```

`\TabLine` prints one line of the font tables. `#1` is the character code of the leftmost character in that line.

```
\def\TabLine #1{%
    \bcount = #1
    \hbox{%
        \vtop{%
            \baselineskip = 14pt
            \hrule height 0pt depth 0pt
            \hbox{Hex}
            \hbox{Octal}
            \hbox{Decimal}
            \hbox{\tt\string\catcode}
            \hbox{{\tt\string\rm}, \dots}
            \hbox{\tt\string\tt}
        }%
        \hskip 8pt
        \OneChar{0}\OneChar{1}\OneChar{2}\OneChar{3}%
        \OneChar{4}\OneChar{5}\OneChar{6}\OneChar{7}
    }
```

The following `\vskip` determines the distance between two consecutive lines.

```
      \vskip 12pt
}
```

Print all four tables. Each table contains 32 characters, therefore all four tables contain 128 characters.

```
\TabLine{0} \TabLine{8} \TabLine{16} \TabLine{24}
\vfill\eject
\TabLine{32} \TabLine{40} \TabLine{48} \TabLine{56}
\vfill\eject
\TabLine{64} \TabLine{72} \TabLine{80} \TabLine{88}
\vfill\eject
\TabLine{96} \TabLine{104} \TabLine{112} \TabLine{120}
\vfill\eject
\end
```

• End of `fonttab.tip` •

16.10 "External" Fonts

"External" fonts are fonts that are not Computer Modern fonts. Adding external fonts to TEX is possible and in this Section I will give you some hints of how to do that.

16.10.1 POSTSCRIPT Fonts

POSTSCRIPT is a high level programming language for the description of graphics and text. It is directly interpreted by certain laser printers, photo typesetters and other output devices. A detailed description can be found in three excellent manuals: POSTSCRIPT (1985a), POSTSCRIPT (1985b) and POSTSCRIPT (1988). To fully understand POSTSCRIPT requires a thorough computer science background. Nevertheless, let me briefly discuss POSTSCRIPT fonts and how they relate to TEX.

16.10.1.1 The Outline Format of POSTSCRIPT Fonts

The main difference between POSTSCRIPT fonts and pixel-file-based fonts, like the Computer Modern fonts, is that all relevant font information of POSTSCRIPT fonts resides in the POSTSCRIPT printer (there are no "POSTSCRIPT-pixel files" somewhere on the computer you are running TEX on). The information stored inside the POSTSCRIPT printer are *not* pixel files. The bitmaps, which are ultimately needed to print POSTSCRIPT font characters, are *computed* by the POSTSCRIPT internal computer from *outline* information stored in the printer.

The major advantage of storing characters in an outline format rather than in bitmap formats is that bitmaps of character in almost any size can be computed, and therefore POSTSCRIPT fonts can be used in almost any size. There are (within very reasonable limits) no restrictions on the font. Particularly there is no need to limit oneself to a set of standard magnifications (like `\magsteps`) as was the case with pixel file based fonts.

16.10.1.2 The POSTSCRIPT Symbol Font

POSTSCRIPT's symbol font is *inadequate* for TEX to do mathematical typesetting. TEX expects many more special characters available in a font for the typesetting of mathematical equations.

The fact that, at least to the date of the writing of this series, there is not an adequate POSTSCRIPT-based math font available is a serious limitations as far as the typesetting with TEX using POSTSCRIPT fonts is concerned. The current solution of mixing POSTSCRIPT fonts and Computer Modern fonts leads to undesirable visual incompatibilities.

16.10.1.3 POSTSCRIPT Drivers

There are various `dvi` to POSTSCRIPT drivers available, which convert a `dvi` file into POSTSCRIPT, and also allow the inclusion of other POSTSCRIPT documents into a TEX generated document[2].

16.10.2 Third Party Fonts

A general distinction into the following two cases can be made when a third party font is used with TEX.

1. The font is METAFONT-based. Because METAFONT generates `tfm` and `gf` files using such a font should pose no problems.
2. The font is *not* METAFONT-based. An example of this is when phototypesetter native fonts is used. In this type of setup one typically faces the following problems:

    (a) The font metric information is not available. It is sometimes difficult to convince the manufacturer of such fonts to provide the required font metric information. After this information is available, it is possible to generate the necessary `tfm` files using `pltotf` (see 17.2.2, p. 316).

---

[2] See Bechtolsheim (1990b) for one example.

(b) The mapping of character codes to characters is different from TEX's default setup. To change those setups is frequently quite a burden. There is another approach that changes the mappings in the `dvi` file; see Bechtolsheim (1989) for details.

(c) Font manufacturers of phototypesetters may not provide pixel files for their fonts which could be used to produce output on a laser printer or on a bit-mapped display. Font emulation can be used to solve this problem; see Bechtolsheim (1989) for details.

16.11 Summary

In this chapter we learned:

- Font substitution macros as well as macros to load fonts on demand were defined.
- Fonts can be organized in groups by typeface, independently of that by size. It is rather easy to load fonts on demand and substitute fonts, if a certain font is not available.
- The space between words is handled as interline glue by TEX. The currently active font is responsible for the amount of interline glue. The value can be changed by setting **`\spaceskip`**.
- The space after punctuation symbols is normally extended. The currently active font is responsible for the size of the extended space, unless **`\xspaceskip`** has been set.
- Two different "spacing modes" are available: french spacing and non-french spacing.
- Executing METAFONT generates one `tfm` file for every font at a specific size. METAFONT also generates **`gf`** files (different files for different magnifications of fonts and different resolutions in the output devices used). **`gf`** files are frequently converted into **`pk`** files which are more compact than **`gf`** files.
- The numerical pixel file extension in **`gf`** and **`pk`** files discriminates among different magnifications and resolutions of a font.
- TEX knows kerning and handles ligatures automatically. The necessary information is taken from `tfm` files.
- Characters are treated internally by TEX as numbers, based on the ASCII character code. Typically fonts in TEX have 128 characters, but 256 characters are also possible.
- The working principle of raster output devices (e.g., laser printers) is based on a pixel matrix, where each pixel can be turned on and off individually to compose a picture.
- Accents in text and in math mode are treated differently in TEX. This chapter discussed text accents and showed the definitions of some accent macros of the plain format.

- TEX's capabilities to underline text are limited.
- Macros to present font tables were discussed.
- External fonts (non-Computer Modern fonts) such as POSTSCRIPT fonts were discussed briefly.

17
In and Around TEX

The purpose of this chapter is to discuss how TEX interfaces with "the rest of the world." I discuss the various versions of the TEX program, what METAFONT is and how it works together with TEX. You will also find a discussion of the WEB programming system that was used to develop TEX and METAFONT. We will also discuss those problems that occur in the production of documents that TEX does not solve conveniently. At the end of this Chapter you find an overview of all TEX-related file extensions.

17.1 TEX Program Variants

A variety of different instances of the TEX program exist normally on any machine. For instance, when using the command `tex` the TEX program normally starts-up with the plain format pre-loaded. Or when you enter the command `latex` then the TEX program with the LATEX format is started. The question addressed here is how can you start the TEX program *without* the plain format pre-loaded or any other format pre-loaded.

Remember that unless a macro package is loaded, only the primitives of TEX are available; see 21.2, p. III-152, for a discussion of primitives. Remember also that the term *format* was defined as a set of macro definitions, which are suitable for typesetting at least one class of documents using TEX.

17.1.1 Explaining `initex` and `virtex`, `plain.tex`, `\dump`

The TEX program comes in two versions, `initex` and `virtex`. `initex` is used to generate format files which can be loaded fast. These format files are used with `virtex`. You will probably wonder at this point in time what happened to the "ordinary" `tex` command which is not listed with `initex` and `virtex`. Also that question will be answered shortly.

1. **initex**. The **initex** version of the TEX program is for the generation of **fmt** files. The abbreviation "**fmt**" stands for format. The purpose of **fmt** files is simply to save time when loading long macro files.

   Assume you are using the plain format of TEX. The macro definitions of the plain format are normally stored in a file called **plain.tex**. It would be very time consuming if TEX had to process all of the plain format macro definitions at the beginning of each job. Therefore with the help of **initex** the following steps are executed:

   (a) Start **initex** to read in **plain.tex**.
   (b) Generate a **fmt** file (called **plain.fmt**). This is done by entering **\dump** after **initex** finished reading **plain.tex**. **Initex** will automatically halt after the **fmt** file is written.

   The generated **fmt** will be read in by **virtex**. Observe that no document was processed or any output was generated so far.
2. **virtex**. The **virtex** version of the TEX program can read **fmt** files. Two steps are executed by **virtex** to process a document.

   (a) Read in the **fmt** file, for instance, **plain.fmt**.
   (b) Read in the document source file.

   The command line to invoke **virtex**, telling **virtex** to load **plain.fmt** and to process some file **text.tex**, typically looks as follows: **virtex &plain test**. Observe the use of the '**&**' to indicate to **virtex** to load a **fmt** file rather than an ordinary TEX source file.
3. In certain operating systems (UNIX is an example of such a system), the process of speeding up the loading of formats can be pushed even further. For that purpose the following steps are executed:

   (a) **virtex** is started by loading the desired **fmt** file, for instance, **virtex &plain** (no document source file is specified).
   (b) After **virtex** finished reading the **fmt** file, a core dump of the currently executing **virtex** is triggered and written to a file. This core dump contains the **virtex** program and the format file in the memory allocated by **virtex**.
   (c) Now the core dump file is converted into an executable program that later can be started up directly. Thus a version of **virtex** with the format file already loaded has been generated. In the case of UNIX, the program which converts a core dump into an executable program is called **undump**.

17.1.2 Commands tex, latex, amstex

The ordinary user of TEX will normally use one of the following three commands to invoke TEX:

1. **tex**: TEX with the plain format pre-loaded is started.
2. **latex**: TEX with the LATEX format pre-loaded is started.
3. **amstex**: TEX with the AMS-TEX format pre-loaded is started.

In all of these cases, it would be more accurate to say that **virtex** is started. To be even more precise; one of the following two alternatives apply:

1. The commands are nothing else than abbreviations for executing **virtex** and loading the proper format file; for instance, the command **tex** is an abbreviation for **virtex &plain**.
2. The commands are the names of programs generated as a result of executing step 3 (the "undump" step) of the list of the previous Subsection.

17.1.3 Performance Comparison of initex versus virtex

Generating a **fmt** file with **initex**, which is then read in by **virtex**, is done for efficiency only. In other words one could just use **initex** and make each TEX source file be processed with the plain format by inserting **\input plain.tex** at the beginning.

The differences in performance between using **initex** and using **virtex** with **plain.fmt** are significant: in an experiment I performed on a VAX 11/780 (an "old machine" by today's standard) running UNIX BSD 4.2, **initex** took 80 seconds to read in **plain.tex** (and 10 seconds to dump **plain.fmt**). **Virtex** on the other hand took only 11 seconds to load **plain.fmt**. The **initex** step is not repeated unless there is a change in the source of the format, of course.

Here is a brief summary of the differences between **initex** and **virtex**:

1. Two versions of the TEX program, **initex** and **virtex**, exist because of performance considerations.
2. Hyphenation patterns which control TEX's hyphenation can only be read in by **initex** and they can be only used by **virtex**. In other words, the hyphenation pattern cannot be changed when regular typesetting is done. See 12.7, p. 137, for more details on hyphenation.
3. **initex** can write, but not read **fmt** files. The opposite is true for **virtex**: it can read **fmt** files, but not write them.

17.2 METAFONT

TEX's companion program is METAFONT. METAFONT is a program to generate fonts. METAFONT's input contains a geometric description of the characters of a font. Characters are described in terms of pen movements along splines, straight lines, and circles. This input is stored in a METAFONT source file.

The METAFONT program takes the information from the METAFONT source file and generates a `tfm` and a `gf` file. The major advantages of generating fonts based on a mathematical description of the shapes of each character are as follows:

1. It is possible to generate `gf` files for printers of almost any resolution.
2. Different printer engines requiring slightly different `gf` files can be accommodated easily (this refers to the issue of write-black versus white-write engines).
3. It is easy to modify all characters in a font by some simple changes (for instance, to generate an italicized version from a non-italicized METAFONT source based font).
4. The same METAFONT source file can be used to generate a font at different sizes.

For details on METAFONT see Knuth (1986c). See also 16.3.3, p. 287.

17.2.1 The Interfacing of TEX and METAFONT

Now let me discuss how TEX and METAFONT interface. TEX will read in the `tfm` files generated by METAFONT. The `tfm` files provide TEX with the necessary font information, in particular with the sizes of characters. See 15.2.7, p. 238, for details on `tfm` files. The driver used to print a document will read in the `gf` files generated by METAFONT to generate the pixel patterns for the characters used to print the document; see 16.3.3, p. 287, for a comparison of the various pixel file types.

17.2.2 `tfm` File-Related Utilities, `tftopl`, `pltotf`

There are two `tfm` file-related utilities:

1. `tftopl`. This program converts a `tfm` file into a property list file. *Property list files* are textual representations of `tfm` files. This way a `tfm` file can be examined easily.
2. `pltotf`. This program performs the inverse function of `tftopl`: it converts property list files back into `tfm` files. This program is useful when a `tfm` file is generated by a program other than METAFONT. This other program can generate a property list file which can be easily read and checked. The generation of the `tfm` file is then left up to `pltotf`, which will also perform some checks to insure the correctness of the `tfm` file.

17.3 The `WEB` System

TEX and METAFONT have been programmed using the "programming language" `WEB`. (I use quotation marks because `WEB` is actually more than a programming language: it is a programming philosophy (called literate programming) combining programming and documentation). What follows is a very simplified explanation of the `WEB` system. See WEB (1988) for the document coming with the `WEB` software. An excellent description of `WEB` can be found in Sewell (1989).

The `WEB` system has many different goals of which we will discuss two now. First it is to provide a programming system that integrates documentation and programming. This is achieved by entering a PASCAL-based program together with a TEX-based documentation into a `WEB` source file. Second it is to support porting systems to different operating systems. This is done through change files, which are discussed shortly.

17.3.1 Combining Documentation and Programming, `weave` and `tangle`

Programming and documentation are combined in `WEB` source files as follows: the `WEB` programmer mixes TEX input for the documentation with PASCAL code for the programming. The `WEB` source file will then be processed by the following two programs:

1. **`tangle`**. The `tangle` program will take the `WEB` source file, reorganize it, perform certain macro expansions and remove the documentation. The result is a PASCAL program, which can be compiled with a regular PASCAL compiler. TEX is usually C-based these days and so versions of `tangle` generating C code from `WEB` source code prevail today.
2. **`weave`**. The `weave` program will take the `WEB` source file and convert all of its program parts to TEX code; the TEX-based documentation already contained in the `WEB` source file is also modified, but only very little. An index is also generated. The result of these steps is a TEX source file containing documentation and a program that when processed by TEX, results in a typeset version of the program listing.

If you are interested in real samples of `WEB` program listings see Knuth (1986b), Knuth (1986d) and Sewell (1989).

17.3.2 Porting `WEB` Programs, Change Files

Another `WEB` system idea is the concept of *change files*. These files allow you to port a `WEB` program to another operating system efficiently. As the term *change file* suggests, a change file contains changes to be made. In the case of the WEB system change files are intended to contain all necessary changes to port a WEB program to another operating system.

Assume, for instance, that you have to change the way the TEX program interfaces with the operating system to read in the time and date (TEX has parameters `\year`, `\month`, `\day` and `\time` which must be initialized by calling the operating system). If your operating system does *not* have a function that returns the number of minutes after midnight (what needs to be loaded into `\time`), then you need to change TEX's source code. Assume here that what your operating system has is a system call `time()`, which returns the current time encoded as the number of seconds since January 1, 1952. You now must connect your version of the TEX program to read this number and to compute from it the number of minutes after midnight, which then is loaded into the register `\time`. Obviously all the other time related parameters need to be connected to the operating system in a similar fashion.

The hacker's old approach was simply to go into the WEB source file and *change* the source file directly. Each time a new version of a program is released, those changes have to be reapplied to the new source file, an error-prone and tedious process. It is more efficient to store all changes which depend on the operating system in a separate file, the change file. Both programs, `weave` and `tangle`, read the WEB source and modify it according to the information from the change file.

The idea of using a change file to port TEX (and of course METAFONT) to a variety of operating systems is one of the reasons why TEX and METAFONT were ported to so many machines. Once a change file is developed, it is almost always possible to use this change file for every new version of a WEB source file, because normally changes in the WEB source file and changes introduced through the change file are disjoint. In other words, as a program develops over time, the change file needs to be adjusted only rarely.

17.4 Environment Variables and Logical Names

We will now discuss the question of how TEX locates various input files. For this Section define *path* as an *ordered list of directories.* TEX will traverse this list in the given order to locate a file.

17.4.1 Listing the Relevant Environment Variables

In UNIX such paths are normally defined by environment variables. Here is a list of the four environment variables as they are commonly used by UNIX implementations of TEX.

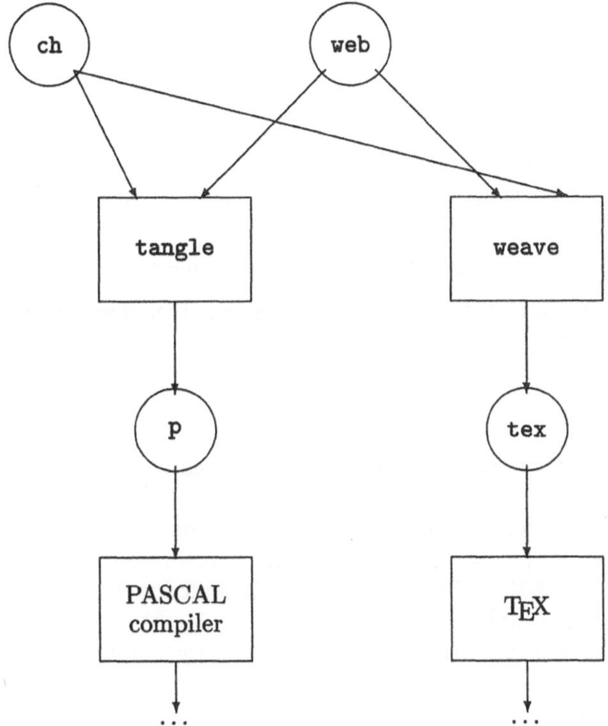

Figure 17.1. The WEB system.

1. **TEXINPUTS**. This environment variable specifies TEX's path to locate TEX source files, like the main source file or any macro source files.

   TEX will traverse the list of directories in the given order. You must be careful to avoid two TEX source files with the same name but in different directories of the **TEXINPUTS** path. If both directories are contained in the **TEXINPUTS** path, TEX will find the first file in the given list of directories but ignore the second one. No warning message will be printed by TEX.
2. **TEXFONTS**. This environment variable determines the path used by TEX to locate `tfm` files. Drivers frequently use this environment variable to locate the pixel files.
3. **TEXFORMATS**. This environment variable defines TEX's path of locating `fmt` files. This path usually points to one common directory for all users on a specific system plus the current directory for formats defined by the user.

   This path is used by `virtex` to locate any `fmt` files. If a pre-loaded `virtex` is used, then this path is irrelevant. This path also has no meaning to `initex`.
4. **TEXPOOL** determines the path of the TEX program to locate `pool` files. Pool files are part of the **WEB** system; see WEB (1988) for details.

Similar ideas apply to implementations of TeX on other operating systems; for instance, in VMS the corresponding logical names (instead of environment variables) determine the above paths.

17.4.2 Sample Settings of TeX's Environment Variables

Take the following setting of **TEXINPUTS** as an example (the various directories in a path specification are separated by colons from each other; the directory "." stands for the current directory).

```
TEXINPUTS=.:/usr/local/tex/lib/macros:/usr/local/tex/lib/texip
```

This setting instructs TeX to look for any TeX source files in the following directories in the given order:

1. . (the current directory). The current directory must be included in the **TEXINPUTS** path so TeX is able to locate a source file in the directory you are currently working in, in particular to locate the main source file that you are using.
2. **/usr/local/tex/lib/macros**. In our example this is the standard directory for most of the TeX source files. This directory also contains all of LaTeX's style files. If this directory were not included in the **TEXINPUTS** path, you would not be able to use LaTeX.
3. **/usr/local/tex/lib/texip**. This directory contains all the macros presented in this series. I wanted to use these macros in other documents, of course, and therefore I established a separate directory for them.

17.5 What TeX Is Not Designed To Do

There are certain problems which occur in the generation of documents for which TeX is not very suitable (mainly because TeX was not intended to solve those problems in the first place). Let me discuss those problems now and provide you with pointers for how you can solve the related problems.

1. *Diagrams and Graphs.* TeX does *not* support the generation of diagrams and graphs. The only thing you can do in TeX is draw horizontal and vertical lines of arbitrary dimensions; see 5.7, p. I-154, for details.

   LaTeX's picture environment, a completely TeX-internal solution, solves this problem to some degree (all diagrams in this book were printed using the picture environment of LaTeX). It allows the user to draw slanted lines, although there are only a limited number of slopes available. Observe that these slanted lines are generated with a special font. This font does not contain regular characters, but pieces of slanted lines of various slopes (and other "characters"). Long slanted lines are combined from many "short slanted line characters." This approach works perfectly well to generate simple diagrams; still it is what a computer scientist calls a "kludge."

In my view the most promising and flexible solution for including figures and diagrams in TEX documents is to use TEX external programs generating POSTSCRIPT and to import those documents through the driver; see Bechtolsheim (1990b) for a more detailed discussion.

2. *Index.* TEX does *not* support the generation of an index. TEX *does* have the capability of writing information (index terms and page numbers) to files. TEX *can* also read-in information from an external file, like a sorted index file for typesetting.

   But the real issue in indexing is not this writing and reading, but the sorting of the index. Sorting of an index is far from trivial: assume, for instance, that you would like to put the following three terms into an index: **TEX**, `TeX` and `\TeX`. You must be able to instruct the sorting program as how to sort these three closely related terms. Note that the `\bf` of `{\bf \TeX}` in the index file is for the printing of **TEX** that is the `\bf` should be ignored as far as the sorting is concerned. On the other hand, `\bf` by itself might appear in an index.

   Leslie Lamport has provided an index sorting program described in Lamport (1987) for index. I used this program quite successfully for the generation of the index of this series.

3. *Bibliography.* TEX does *not* support the maintenance of a bibliography. One way to solve this program is to use `bibtex`, a program that administers bibliographies in LATEX. See Lamport (1985) for how this program is applied. See Patashnik (1985) for further details.

4. *Spelling checker.* TEX does *not* contain a spelling checker. One typically uses a program like `detex`, which removes any TEX instructions from the copy of a TEX source file, and then passes the file on to a regular spelling checker. This was already discussed in 2.5, p. I-12, item 4.

5. *Printing* `dvi` *files.* TEX's `dvi` file cannot be printed directly without a device driver. By not trying to accommodate every possible printer, i.e., by having a separate program which interfaces with TEX, it is not necessary to change and extend TEX each time a new printer will be used with TEX.

17.6 Utilities

The standard set of TEX utilities that are part of the core distribution of TEX and METAFONT follow. When you acquire a TEX distribution you should make sure that all these utilities are provided. These utilities are normally not distributed to PC users.

1. **`dvitype`**. This program is a kind of master device driver. See 17.7 on the next page for a brief discussion.

2. **tftopl** and **pltotf**. **Tftopl** converts a **tfm** file into a property list, and **pltotf** does the reverse; see 17.2, p. 315, for details.
3. Pixel file-related utilities. As you will see in 16.3.3, p. 287, there are three different types of pixel files: **gf**, **pk** and **pxl** files. The following utilities relating to these files are available:

   (a) **gfread**. See Rokicki (1988) for further details.
   (b) **gftodvi**. See Knuth (1989) for further details.
   (c) **gftype** utility program reads the binary generic-font (**gf**) files that are produced by font compilers such as METAFONT and converts them into symbolic form. This program has two chief purposes: to determine whether a **gf** file is valid or invalid and to serve as an example of a program that reads **gf** files correctly.
   (d) **pktype** prints a textual representation of a **pk** file.
   (e) **pxltype** prints a textual representation of a **pxl** file.
   (f) **gftopxl** converts a **gf** file into a **pxl** file. This is a utility you need only if you did not convert your driver to **gf** or **pk** files.
   (g) **gftopk** converts a **gf** file into a **pk** file.
   (h) **pktopxl** converts a **pk** file into a **pxl** file.
   (i) **pxltopk** converts a **pxl** file into a **pk file**.

4. Other TEX-related utilities:

   (a) **patgen**. This program takes a list of hyphenated words and generates a set of patterns that can be used by the TEX hyphenation algorithm.
   (b) **pooltype**. The **pooltype** program uses **tangle** to convert the output of string pool files to a slightly more symbolic format than may be useful when **tangle**d programs are being debugged. People may want to try transporting this program before tackling TEX itself.
   (c) **profile**. This program inserts instructions into a **WEB** program for *profiling*. Profiling is determining which parts of a program are executed most frequently, and how much time the execution of these parts take. See the description of the **profile** program for further details.

17.7 Device Drivers, DVI Files

Printing a **dvi** file to a printer normally involves two steps (this was already discussed in 2.5, p. I-11, item 2):

1. Convert the **dvi** file into a file in the *printer language*, the language understood by the printer. The conversion is done by a program called a *device driver*.
2. Send the generated file to the printer for printing. A *spooler* serializes the potentially conflicting requests of many users.

An understanding of `dvi` files is important to the TEX user. A `dvi` file is internally subdivided into a *preamble* followed by a *sequence of pages*, which in turn is followed by a *postamble.*

The following type of instructions can be found in a `dvi` file:

1. Instructions indicating the beginning and ending of the preamble, pages, and postamble.
2. Instructions to load and change fonts.
3. Instructions to print some text.
4. Instructions to move horizontally and vertically.
5. Instructions to draw rules.
6. Instructions containing the information contained in `\specials`; see 29.9, p. III-533.

Any dimensions in TEX-generated `dvi` files are based on the unit *scaled points.* If you look at the dimension table in 4.1, p. I-81, then you will find that a scaled point (sp) is a very small unit. All dimensions in `dvi` files are based on integer values in the unit scaled points; no movements of fractions of a scaled point can be given in a `dvi` file.

A laser printer has a *very* coarse resolution compared to the scaled point resolution of a `dvi` file. Therefore, one of the major duties of a device driver is to round dimensions in scaled points from the `dvi` file properly to integer dimensions with respect to the resolution of the printer used. This may sound trivial but it really isn't, particularly if kerning and other "small" details are involved.

The "model device driver" is `dvitype`, a program that reads-in a `dvi` file and print out a textual representation of the `dvi` file. This program computes positions in a way any device driver is expected to do (sadly enough, not all device drivers follow the rules precisely).

In 33.4.5, p. IV-48, the issue of landscape mode versus portrait mode is discussed. This is also an issue that involves a device driver. TEX does not care about the dimensions of a page on which it places text, and therefore has no notion of the orientation of a page.

17.8 Checksums

In order to avoid accidental incompatibilities between `tfm` and any type of pixel files (`gf`, `pk` and `pxl` files), all those files contain checksums. I would like to explain this checksum mechanism briefly.

When METAFONT generates a `tfm` file and a `gf` file (or files) of some font, all those files will have the same checksum stored in them. This checksum is computed by METAFONT in such a way that a change in the METAFONT source code of a font most likely causes a different checksum to be computed when METAFONT is executed.

When TEX accesses a font it reads in the `tfm` file of this font, including its checksum. Then this checksum is written out to the `dvi` file of each document. Thus the driver, when accessing a pixel file, will compare the checksum of the font as found in the `dvi` file with the checksum as stored in the pixel file it accesses. If the driver discovers a mismatch between the two checksums, it will report an error and warn the user that there are inconsistent `tfm` and pixel files being used. The assumption is that somehow the `tfm` file used by TEX and the pixel file used by the driver are not derived from the same METAFONT source code because otherwise their checksums would not be different.

A checksum is by definition non-zero. If a zero checksum is discovered in any of the above files, no checksum testing is performed.

17.9 TEX-Related File Types

This Section provides a list of (hopefully) all file types that are of interest in TEX.

1. **`afm`** files or Adobe font metric files. These files contain the font metric information for POSTSCRIPT fonts. They are regular text files; see 16.10.1, p. 309, for a discussion of using POSTSCRIPT fonts in TEX.
2. **`aux`** files. Auxiliary files of LATEX containing bookkeeping information of the LATEX macro package. Also used by the partial processing macros of this series. See 30.4, p. III-541, for details.
3. **`bbl`** files. Files that are printed by `bibtex` and read-in by LATEX for the typesetting of bibliographies.
4. **`bib`** files. Bibliography data base files that are read-in by `bibtex` and generated by the user.
5. **`ch`** files. `WEB` change files.
6. `dvi` files. TEX's output file containing the typeset document; see Knuth (1986b), sections 583–591, for a detailed description of the `dvi` files as they are generated by TEX. A reprint of those sections can also be found in Fuchs (1982).
7. **`fmt`** files. These files are generated by `initex` and read-in by `virtex`, allowing the efficient start-up of processing a document with TEX; see 17.1.1, p. 313, for details. This file type is described in Knuth (1986b), sections 1299-1329.
8. **`gf`** files. **`gf`** files are generated by METAFONT; see Knuth (1986d), sections 1142–1148, for a description of this file format. A reprint of those sections can also be found in GF (1985). See 16.3.4, p. 288, for details on **`gf`** files.
9. `lof` files. List of figure files of LATEX and the macros of this series (see 30.4, p. III-541).
10. `lot` files. List of tables file of LATEX and the macros of this series (see 30.4, p. III-541).

11. `mf` files. METAFONT source files are regular text files; see 17.2, p. 315, for a discussion of METAFONT.
12. `p`. PASCAL source files. In TeX, these files are normally written by **`tangle`**; see Fig. 17.1, p. 319.
13. `pk` files. These files are normally generated from `gf` files using **`gftopk`**. See Rokicki (1985) for a description of this file type. See 16.3.6, p. 289, for details on `gf` files.
14. `pxl` files. Files of this type are outdated. See Fuchs (1981) for a description of this file type. See 16.3.7, p. 290, for more information on `pxl` files.
15. `tex` files. TeX source files are regular text files.
16. `tfm` files. These files contain font metric information for TeX. See Knuth (1986d), sections 539–547, for a description of this file format. See 15.2.7, p. 238, for more details.
17. `tip` files. All published source files and most of the sample files in this series use the file extension `tip`, which obviously stands for "TeX in Practice."
18. `toc` files. Table of contents files of LaTeX and the macros of this series (see 30.4, p. III-541).
19. `web` files. WEB source files are regular text files.

17.10 The Version Number of the Plain Format Source Code

The plain format's version number is an important but also confusing ingredient to the discussion: there is no clear separation between the version number of the TeX program and the version of the plain format.

Each time a bug in the TeX program is found the number after the decimal point was incremented. When there was a of the Almost Modern Fonts to the Computer Modern fonts, also a version number change (from `1.5` to `2.0` took place), despite the fact that there was no change to the TeX program at this point in time. There were also here and there a few tiny changes to the plain format.

When TeX 3.0 was released the version number of the plain format was also set to 3.0. The changes to `plain.tex` caused by TeX 3.0 are minor only.

17.11 Summary

In this chapter we learned that:

- There are various versions of the TeX program. Usually when invoking the `tex` command, `virtex` will be started up with the plain format pre-loaded. There is `initex` to generate `fmt` files for the fast loading of `virtex`. `initex` is the only version of TEX which allows the reading of hyphenation patterns. `virtex` allows the reading of the `fmt` files that allow the user to preload macro packages fast and efficiently.

- METAFONT is a companion program to TEX for generating fonts. All Computer Modern Fonts are generated using METAFONT.
- TEX and METAFONT were both programmed using the `WEB` system. The `WEB` system supports high quality program documentation and programming. It also supports porting programs to different operating systems efficiently, in particular through the mechanism of change files.
- TEX was ported to a large variety of different machines, from PCs up to very big computers.
- TEX does not support the generation of diagrams and graphs (or at least only in a very restricted sense), and it does not have built-in index sorting facilities or a bibliography reference system.
- Various utilities of TEX allow the conversion of pixel files into various formats and the textual representation of many of the TEX-related files.
- Device drivers convert the `dvi` file produced by TEX into a printer-specific file. A spooler prohibits the simultaneous usage of a printer by more than one user.
- An overview of the various TEX-related file types was presented.

Bibliography

Let me begin this bibliography with some general remarks about books relating to TEX and typesetting.

For books which may improve your knowledge with respect to typesetting and design, I would give the following recommendations: Tufte (1983) is a very interesting book concerning the design of tables. In Chicago (1982) you find a very valuable source concerning styles and any other information pertaining to writing books and many other types of documents. This book provides information about how to do indices, whether or not to capitalize a figure caption, what words to capitalize in a chapter or other title, and so forth. I have tried to follow the standards of this book.

In Lee (1979) you find an excellent book about bookmaking. Strunk (1979) is an old time classic. Concerning typesetting terminology, I find Craig (1978) extremely useful. The explanations in this book are very straightforward and to the point. This book's dedication reads, "Dedicated to the designer who uses phototypesetting without really understanding it." I found this to be true.

Note that Knuth (1986a) is frequently referred to as the TEXbook in this series.

If a reference below contains "`WEB` program," then this means that the documentation is part of a `WEB` program (which by definition contains its own documentation). All of the `WEB` programs cited below belong to a standard TEX distribution. There is now a very nice book you can read about `WEB` which I can highly recommend, because it very eloquently makes the case for "literate programming." See Knuth (1992).

In case you are looking for one of the file format definitions (such as the **pk** file type) be sure to use the references listed below, because some of the file types (for instance, the `dvi` file type) were redefined during the history of TEX.

As far as PostScript is concerned note that there are three books listed below. I added the specification "the red book," the "blue book" or the "green book" to it, because all three books carry distinctive colors and in particular in the early days when there were no alternative PostScript books available one usually referred to these books by their color. The former red PostScript book is now red and white (I personally don't like the new design that much) and that took away some from the colored references to these books.

Anyway, it is time for the bibliography itself now.

Bechtolsheim S (1988) Using the Emacs Editor to Safely Edit TEX Sources, Conference Proceedings, TEX Users Group, Ninth Annual Meeting, Montreal, August 24–24, 1988, TEXniques, Publications for the TEX Community, Number 7.

Bechtolsheim S (1989) A `dvi` File Processing Program, TUGBOAT, 10:3, pp. 329–332.

Bechtolsheim S (1990a) A TEX Font Catalogue, Integrated Computer Software, Inc., West Lafayette, IN.

Bechtolsheim S (1990b) TEXPS, A TEX POSTSCRIPT Software Package, Integrated Computer Software, Inc., West Lafayette, IN.

Bechtolsheim S (1990c) A Universal TEX Preprocessor and `make` Related Utility, Integrated Computer Software, Inc., West Lafayette.

Chicago Manual of Style (1982), Thirteenth Edition, The University of Chicago Press, Chicago

Craig J (1978) Phototypesetting: A Design Manual, Watson-Guptill Publications, New York.

Victor Eijkhout (1992) TEX by Topic, a TEXnician's Reference, Addison-Wesley Publishing Company, Inc., Reading, MA.

Fuchs DR (1981) The Format of `PXL` Files, TUGBOAT 2:3, pp. 8–12.

Fuchs DR (1982) The Format of TEX's `dvi` Files, TUGBOAT 3:2, pp. 14–19.

Guidelines (1990) Guidelines for the Preparation of a Typeset Camera-Ready Manuscript, Springer-Verlag, New York.

Herwijnen Eric van (1990) Practical SGML, Kluwer Academic Publishers, Dordrecht, Netherlands.

Kabelschacht A (1987) **`\expandafter`** vs. `\let` and **`\def`** in Conditionals and a Generalization of PLAIN's `\loop`, TUGBOAT 8:2, pp. 184–185.

Knuth DE (1983) The `PROFILE` Processor, Version 1.1, November 1983, `WEB` program.

Knuth DE (1986a) The TEX Book, Addison-Wesley Publishing Company, Inc., Reading, MA.

Knuth DE (1986b) TEX: The Program, Addison-Wesley Publishing Company, Inc., Reading, MA.

Knuth DE (1986c) The METAFONT Book, Addison-Wesley Publishing Company, Inc., Reading, MA.

Knuth DE (1986d) The Metafont Program, Addison-Wesley Publishing Company, Inc., Reading, MA.

Knuth DE (1986e) Computer Modern Fonts, Addison-Wesley Publishing Company, Inc., Reading, MA.

Knuth DE (1988) The `WEB` System of Structured Documentation, `WEB` program (**`tangle`**, **`weave`**) (documentation is part of the software).

Knuth DE (1989) The `GFtoDVI` Processor, Version 3.0, a `WEB` program.

Knuth DE (1990) The New Versions of TEX and METAFONT, TUGBOAT, 10:3, pp. 325–328.

Lamport L (1985) The LATEX Manual, Addison-Wesley Publishing Company, Inc., Reading, MA.

Lamport L (1987) *Makeindex*: an Index Processor for LATEX. Source code of document comes with the **`makeindex`** software.

Lee M (1979) Bookmaking: the illustrated guide to the design/production/editing, R. R. Bowker, New York.

Patashnik O (1985), BIBTEXing, Version of May 23, 1985 (this documentation comes with the BIBTEX software).

Platt C (1985) Macros for Two-Column Format, TUGBOAT 6:1, pp. 29–30.

POSTSCRIPT (1985a) POSTSCRIPT Language, Tutorial and Cookbook, Adobe Systems, Inc., Addison-Wesley Publishing Company, Inc., Reading, MA. The *blue* book.

POSTSCRIPT (1985b) POSTSCRIPT Language, Reference Manual, 2nd Edition, Adobe Systems, Inc., Addison-Wesley Publishing Company, Inc., Reading, MA. The *red* book.

POSTSCRIPT (1988) POSTSCRIPT Language, Language Program Design, Adobe Systems, Inc., Addison-Wesley Publishing Company, Inc., Reading, MA. The *green* book.

Rokicki T (1985) Packed (PK) Font File Format, TUGBOAT 6:1, pp. 115–120.

Rokicki T (1988) The `GFread` Processor, Version 1.1, July 11, 1988, `WEB` program.

Sewell W (1989) Weaving a Program, Literate Programming in `WEB`, Van Nostrand Reinhold, New York.

Shakespeare W (1605) King Lear, Edited by Alfred Harbage, Penguin Books, New York, revised edition 1970.

Solomon D (1990a) Output Routines: Examples and Techniques. Part I: Introduction and Examples, TUGBOAT 11:1, pp. 69–84.

Solomon D (1990b) Output Routines: Examples and Techniques. Part II: OTR Techniques, TUGBOAT 11:2, pp. 212–249.

Spivak M (1986) The Joy of TEX, American Mathematical Society, Providence, RI.

Spivak M (1989) LAMS-TEX The Synthesis, TEXplorators Corporation, Houston, TX.

Stallmann R (1986) GNU Emacs Manual, Free Software Foundation, Cambridge, MA.

Strunk W, Jr. and White EB (1979) The Elements of Style, Macmillan Publishing Co., Inc., New York, NY.

Tufte ER (1983) The Visual Display of Quantitative Information, Graphics Press, Cheshire, CT.

Webster (1985) Webster's Standard American Style Manual, Merriam-Webster, Inc., Springfield, MA.

Index

Each volume contains a comprehensive index. The index is followed by a separate index of all published source code files. Primitives in the index are marked by an asterisk, e.g., *\par.

Source Code File Index

This index is an index of all the *published* source code files in this series. Most macro source code files belong to the **texip** format (defined in 31.3, p. III-612). If you use this format no special action is necessary to use most of the published macro source files of this series. Those files which do *not* belong to the **texip** format are marked by an asterisk (*).